Preparing Data for Analysis with JMP®

Robert Carver

sas.com/books

Contents

About This Book

What Does This Book Cover?

In a 2008 interview, Google's chief economist, Hal Varian, remarked:

> *I keep saying the sexy job in the next ten years will be statisticians. People think I'm joking, but who would've guessed that computer engineers would've been the sexy job of the 1990s? The ability to take data—to be able to understand it, to process it, to extract value from it, to visualize it, to communicate it—that's going to be a hugely important skill in the next decades...* [1]

Perhaps the very least attractive aspect of the "sexy job" is the work involved in assembling, reconciling, tidying, cleaning, and otherwise preparing data from various sources *before* the serious work of processing, extracting value, visualizing, and communicating. Although data preparation typically consumes an enormous share of the time in most projects, it receives comparably little attention in the data analytics literature.

It is as if data preparation is a dark art or a nasty family secret, widely acknowledged but not spoken about in polite company. This book is all about using the extensive capabilities of JMP to facilitate and regularize the phases of preparing data for analysis.

This book is entirely and exclusively about the stages that precede the actual analysis in a statistical investigation. It covers methods for extracting data from various sources and in different formats and converting them to JMP data tables. Because so many projects call for merging multiple data tables, we see how the powerful JMP Query Builder for Tables facilitates such operations, enabling the analyst to manage data consolidation at scale and at relatively high speed.

As practitioners know all too well, once the data are all in one place, the real adventure begins. JMP can also speed the work of "rounding up the usual suspects" of dirty data: missing observations, outliers, mismatched key fields, and implausibly perfect relationships. After identifying issues, we have an array of alternatives for resolving, mitigating, and managing them. Finally, the book also covers options for communicating data and results to non-users of JMP.

JMP typically offers multiple options to tackle a given issue, and the book presents both the alternatives and a sense of what to use when. For techniques that are out of the mainstream (for example, Principal Components Analysis for data reduction and mitigation of missing cases), chapters provide both introductions to the methods and reliable references to other sources.

Anticipating that some preparatory work might need to be done repeatedly, another recurring theme is reproducibility. Whether a reader is conversant with JMP Scripting Language, users can create a reproducible audit trail by pointing and clicking. The processes illustrated in the book can be preserved by simply saving the scripts that are being written in the background every step of the way.

Oddly enough for a book that prominently features statistical software, there is scarcely any coverage of analytic techniques here, except insofar as a technique helps identify or repair a data problem. The analytic techniques are always in mind, because data preparation must be informed by the varying requirements of different techniques. However, the analytic platforms remain off-stage, as it were.

The book also does not cover JMP fundamentals. There are numerous books, videos, and other training materials for the new user, and readers who are just encountering JMP for the first time are better served to start elsewhere.

Is This Book for You?

If you are a practitioner working with messy data from multiple sources, this book can help. In particular, this book is for practitioners with access to raw data and a pressing need to, in Varian's words, "understand it, to process it, to extract value from it, to visualize it, to communicate it."

What Are the Prerequisites for This Book?

You should have some grounding in statistical methods, and have a working acquaintance with JMP. Some of the features illustrated in the book are available only in JMP Pro, so ideally readers have a JMP Pro license.

What Should You Know about the Examples?

This book includes tutorials for you to follow to gain hands-on experience with JMP. After working through the varied examples within the chapters, you should be able to apply concepts and techniques to your own data. The topics for examples are drawn from various fields. If you need to learn data preparation skills for your own work, you are probably aware that analysis projects require a combination of subject-area knowledge, statistical thinking, and familiarity with data structures. Because each reader brings different domain expertise, the examples are intended to be accessible to most readers regardless of professional background.

Some use data tables that are included with your JMP installation Sample Data Library. Others come from public domain sources, and all of the data tables shown in the book are available for download at http://support.sas.com/publishing/authors/carver.html.

Software Used to Develop the Book's Content

This book was prepared using JMP Pro 13. Readers with earlier versions will find that some menus have changed and that some functionality is not available to them.

Example Data

As noted earlier you can access the example code and data for this book by linking to its author page at http://support.sas.com/publishing/authors/carver.html.

Output and Graphics

Most of the images in the book were captured from within JMP Pro 13, and a few others from websites. Printed editions of the book are in black-and-white, which plainly does not convey the effective use of color produced by JMP in some reports. Electronic editions do render graphics in color.

Where Are the Exercise Solutions?

Most chapters include extended hands-on examples with detailed instructions. To get the most from the book, please follow along and actually work through the exercises and examples. There are no "homework" exercises at the end of chapters, because the goal of the illustrations is to help readers with their own projects and data. Hence, the solutions appear right within the chapter, and your screen should match up with the many screen captures provided in the text.

Acknowledgments

I am very grateful to SAS Press for the invitation and encouragement to write this book, and particularly to Sian Roberts and Julie Platt for their patience throughout the process. Kathy Underwood speedily and expertly copy-edited the entire manuscript. Some chapters would truly have not materialized without expert consultations and suggestions from members of the JMP development team including Brady Brady, Michael Hecht, Eric Hill, Don McCormack, Heman Robinson, and Russ Wolfinger.

The JMP Early Adopter program was indispensable in preparing the manuscript. Thanks to Jeff Perkinson and Daniel Valente for access and updates. JMP Academic Ambassadors Curt Hinrichs, Mia Stevens, and Volker Kraft all assisted in more ways than they can imagine.

Eric Hill, Mike Vorburger, and Richard Zink caught errors, rescued me from rabbit holes, and recommended numerous improvements to earlier drafts. Thank you for your invaluable service as reviewers.

To friends and colleagues outside and inside the "JMPiverse" for data, for perspective, and for pedagogical recommendations: Max Harpers of GroupLens and the University of Minnesota, Professor Nick Horton of Amherst College, Rob Lievense of Perrigo, and Professor Michael Salé of Stonehill College.

Finally, to my wife (and sometimes technical consultant on database matters) Donna, from whom I've stolen too many Sundays together to write this book—Thank you, sweetie.

We Want to Hear from You

SAS Press books are written *by* SAS Users *for* SAS Users. We welcome your participation in their development and your feedback on SAS Press books that you are using. Please visit https://support.sas.com/publishing to do the following:

- Sign up to review a book
- Recommend a topic
- Request information about how to become a SAS Press author
- Provide feedback on a book

Do you have questions about a SAS Press book that you are reading? Contact the author through saspress@sas.com or https://support.sas.com/author_feedback.

SAS has many resources to help you find answers and expand your knowledge. If you need additional help, see our list of resources: https://support.sas.com/publishing.

---

[1] McKinsey & Company. 2009. "Hal Varian on how the Web challenges managers." Available at http://www.mckinsey.com/industries/high-tech/our-insights/hal-varian-on-how-the-web-challenges-managers.

About The Author

 Robert Carver is Professor of Business Administration at Stonehill College in Easton, Massachusetts, and Senior Lecturer at the International Business School at Brandeis University in Waltham, Massachusetts. At both institutions, he teaches courses on business analytics in addition to general management courses. He is the author of *Practical Data Analysis with JMP®, Second Edition.* His primary research interest is statistics education, and he is an Associate Editor for the Journal of Statistics Education. A JMP user since 2006, Carver holds an A.B. in political science from Amherst College in Amherst, Massachusetts, and an M.P.P. and Ph.D. in public policy from the University of Michigan at Ann Arbor.

Learn more about this author by visiting his author page at http://support.sas.com/publishing/authors/carver.html. There you can download free book excerpts, access example code and data, read the latest reviews, get updates, and more.

Chapter 1: Data Management in the Analytics Process

Introduction

Although reliable estimates are difficult to come by, there seems to be consensus that data preparation—locating, assembling, reconciling, merging, cleaning, and so on—consumes something like 80% of the time required for a statistical project (Press 2016). In comparison to the literature about building statistical models and performing analysis, there are relatively few books written on the topic of data preparation. (Some good examples include McCallum 2013, Osborne 2013, and Svolba, 2006.) This book addresses just that—the unglamorous, time-consuming, laborious, and sometimes dreaded "dirty work" of statistical investigations. This work is variously known as data wrangling, data cleaning, the "janitorial work," or simply data management. Equally important, this book is about data management using JMP, so that JMP users can do nearly all of the wrangling tasks using one software environment, without having to hop off and onto different platforms.

This first chapter places data management and preparation in the context of a larger investigative process. In addition, it introduces some over-arching themes and assumptions that run through the eleven chapters to follow. As a starting point, let's understand that the work of data management occurs within a process and often within an organizational context.

A Continuous Process

Successful analyses require a disciplined process similar to the one shown here in Figure 1.1 (SAS, 2016). In such a process, a question gives rise to fact-gathering, analysis, and implementation of a solution. The solution, in turn, is monitored and gives rise to further questions. Though process is often portrayed as circular, this particular depiction has the added appeal of connoting an infinite loop with data playing a central role. However, as we'll see in the pages that follow, the step identified as "Prepare" is not drawn remotely to scale. That's the bit that reportedly can require 80% of the time in a project, and will consume an even larger proportion of the pages in this book.

Figure 1.1: One Model of the Analytics Life Cycle (SAS)

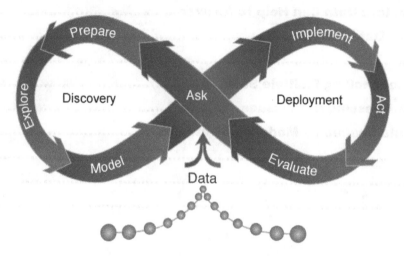

For the most part, this book breaks down that single step into component parts, explaining obstacles that arise and offering techniques to address them. JMP is particularly well suited to expediting the smaller steps that make up the work of preparation. In addition, we'll step into the exploratory stage in later chapters, cycling back as exploration reveals the need for additional pre-processing and preparation. In any event, though, the process properly starts with questions that can be addressed with data.

Asking Questions That Data Can Help to Answer

One underlying assumption in this book is that we undertake statistical investigations because someone has questions (a) that might be answered via empirical investigation and (b) whose answer potentially has measurable value to an organization, society, or an individual. In other words, we'll assume that the questions *matter* to someone. That is, there are benefits to be reaped from finding answers. This also implies that those asking the questions are likely to be attuned to the costs of securing the data and performing relevant analysis.

If you are reading this book, you probably have questions that are important to your work or areas of interest. Which environmental hazards have the greatest impact on respiratory health? Which customers will most likely prefer product A over product B? Do tax cuts encourage employers to create new jobs?

Although the questions driving a study often arise within an organization, the data most germane to the question might reside inside or outside the organization. In this discussion, let's refer to the organization asking the questions as the client. The client might have some of the needed data within its own files and databases (you hope the data are in digital form). Other data might be freely available in the public domain, while other data might be available for a price. Yet other data could be proprietary belonging to competitors, and still other information might not yet have been gathered or curated by anyone.

Early in the process, the analyst or analysis team need to determine the specific data that will serve to address the questions posed or build the models that will have value. This requires domain knowledge, an understanding of what is organizationally feasible and an awareness of what data are available. Perhaps hardest of all, it can require a knowledge of what the team does not know. For some problems, the point of the analysis is *feature selection*—that is, identifying the variables that matter most, that explain or predict a dependent variable or outcome. These problems cannot be resolved through software, but the search for relevant data certainly can be either facilitated or impeded by the software.

Sourcing Relevant Data

Once the analysts have initial ideas about the types of data to obtain, the challenges of locating pertinent data can be considerable. As noted, the organization's own data stores might not have exactly what is needed. Sometimes, organizational dynamics or legal or ethical considerations can impede access.

If we're lucky, external data reside in the public domain and are easily accessible. In other cases, the data are proprietary and belong to a competitor or to an entity that will make them available at a price. Worst of all, sometimes the data simply do not exist in any accessible form, so the analytics team will either need to gather data or use proxy variables, "near enough," if you will, to represent the constructs of interest in a study.

This book has little to say about the many hurdles of data procurement. Chapter 3 discusses some of them, but in general this is an area where domain knowledge is key. The analytics team and the client organization need to know where to look for relevant data.

Reproducibility

As you begin to extract data, it is quite important to document the process in detail for the sake of reproducibility. Some projects are one-time, unique tasks, but for others there is value in being able to reproduce all of the steps taken. Where there are considerations of intellectual property rights, this is critical. If you simply want the ability to audit the task for completeness and critical evaluation, a full record and audit trail is indispensable.

The individual most likely to want to reproduce the process in the future is *you*. Whether others will one day retrace your footsteps, it is altogether likely that you or your team will go back to fetch additional variables, or to build another project on the foundation of this one. So, as the saying goes, be kind to your future self and document (Gandrud 2015, p. 7).

As we'll see throughout the book, documentation is quite straightforward within a JMP session. What's more, we can document selectively preserving results and scripts that we ultimately judge worth saving without having to enshrine every error and false start.

Combining and Reconciling Multiple Sources

Many data modeling projects call for data from multiple sources. In addition to locating all of the variables that you might use, analysts often confront the need to reconcile disparities across data sources. Nonstandard abbreviations or coding schemes, different representations of times and dates, and varying units of measurement abound. Before all of the data can be assembled into a single, well-organized table suitable for analysis, the differences need to be ironed out.

Because the irregularities can present themselves at different stages of the preparation phase, they are discussed in several chapters. Chapter 5 through 8 cover many of the aspects of combining and reconciling data.

Identifying and Addressing Data Issues

Once data sources have been identified and targeted, you must consider issues of data integrity. It is the rare data source that supplies complete, accurate, timely raw tabulated data. For many analytical purposes, we'll want (or require) data tables that are in third normal form – variables in columns and observations in rows, but some data sources won't be organized that way. Tables will have missing cell, sparse arrays (mostly zeros), or erroneous data values. There will be outliers, skewed distributions, non-linear relationships, and so on.

The later chapters devote considerable attention to (a) ways of detecting such data problems and (b) addressing them. Here again, JMP has extensive functionality to ease the way forward. Note also, that our goal is to "address" the issues. That is, we cannot always resolve or eliminate the problems. There are generally ways to mitigate data issues when they cannot be directly cured. These are among the matters taken up in Chapters 9 through 11.

Data Requirements Shaped by Modeling Strategies

The analysis plan for a project influences, or should influence, the data management and preparation activities. Modeling methodologies have their own requirements for the organization of a data table, for units of analysis, and for data types. As a simple and familiar example, models built on paired observations will expect to find data pairs in separate columns.

Hence, you need to be constantly mindful of the stages to follow when preparing a set of data for analysis. Data preparation happens within the context of the full investigative cycle for a reason, and that goes beyond variable selection. Chapters 7 and Chapters 9 through 11 touch on these issues.

Plan of the Book

The investigative process is neither linear nor purely sequential, though we depict it as a logical sequence. That notwithstanding, analysts regularly need to loop back to an earlier phase along the way to a successful conclusion.

In a similar way, books do need to present information in sequence. But readers are free to depart from the plan, to iterate, and to bypass materials that are already familiar. Still, it's helpful to have a mental map of the plan before deviating from it.

I've developed the book in three main sections. Part I consists of Chapters 1 through 3. This part builds a foundation that explains the investigative cycle, introduces common methods used to organize raw data, and reviews the array of common challenges arising in different data sources.

Part II gets down to the "how" of acquiring and structuring a collection of variables for building models. Chapters 4 through 8 present some background concepts and theory, and work through several examples of importing data into JMP from foreign sources and then combining disparate elements into single JMP data tables. Running through Part II is a comprehensive illustrative example about the competitive success of world nations in the summer Olympic Games.

Lastly, Part III (Chapters 9 through 12) covers approaches to some of the thorniest data preparation problems: how to detect problematic data, how to deal with missing data, and how to transform and otherwise modify variables for analysis. Chapter 12 reverses, in a sense, the work of data acquisition by demonstrating ways to share or export data and results to platforms other than JMP.

Conclusion

The main take-away from this chapter is that data management is a large and complex piece of a larger and more complex process. It might not be the most glamorous part of modeling and data analytics, but it is as essential to success as proper soil preparation is to abundant crop yields in agriculture. This chapter has described the process with broad strokes and outlined the tasks the lie ahead.

Before we begin the search for data to wrangle and manage, it's useful to understand some things about the underlying structure of data that we might find. The next chapter reviews and explains the most common ways of organizing and representing raw data.

References

Asay, Matt. 2016 "NoSQL keeps rising, but relational databases still dominate big data." *TechRepublic.com*, April 5, 2016. Available at http://www.techrepublic.com/article/nosql-keeps-rising-but-relational-databases-still-dominate-big-data/.

Carver, Robert., Michelle Everson, John Gabrosek, Nicholas Horton, Robin Lock, Megan Mocko, Allan Rossman, Ginger Holmes Rowell, Paul Velleman, Jeffrey Witmer, and Beverly Wood. 2016. *Guidelines for Assessment and Instruction in Statistics Education (GAISE) College Report 2016*. Alexandria VA: American Statistical Association.

Gandrud, Christopher. 2015. *Reproducible Research with R and RStudio, 2ⁿᵈ Edition.* Boca Raton, FL: CRC Press.

McCallum, Q. Ethan, ed. 2013. *Bad Data Handbook.* Sebastopol, CA: O'Reilly Media.

Osborne, Jason W. 2013. *Best Practices in Data Cleaning: A Complete Guide to Everything You Need to Do Before and After Collecting Your Data.* Thousand Oaks CA: Sage.

Press, Gil. 2016. "Cleaning Big Data: Most Time-Consuming, Least Enjoyable Data Science Task, Survey Says." *Forbes*, March 23, 2016. Available online at http://www.forbes.com/sites/gilpress/2016/03/23/data-preparation-most-time-consuming-least-enjoyable-data-science-task-survey-says/#4fe4cf1e7f75.

SAS Institute Inc. 2016. SAS Institute white paper. "Managing the Analytical Life Cycle for Decisions at Scale: How to Go From Data to Decisions as Quickly as Possible." Cary, NC: SAS Institute Inc.

Svolba, Gerhard. 2006. *Data Preparation for Analytics Using SAS.* Cary, NC: SAS Institute Inc.

Tintle, Nathan., Beth L. Chance, George W. Cobb, Allan J. Rossman, Soma Roy, and Jill VanderStoep. 2014. *Introduction to Statistical Investigations.* Hoboken, NJ, Wiley.

Wild, C. J., and M. Pfannkuch. 1999. "Statistical Thinking in Empirical Enquiry." *International Statistical Review*, 67(3), 223-265.

Chapter 2: Data Management Foundations

Introduction

Data analysis and modeling requires raw data in digital form. At some point in the investigative process, a person or an automated sensor captured data from a real world event or entity and entered and recorded it digitally. People make decisions about what to capture, when to capture, how much to capture and how best to organize data, and these decisions can be based on numerous considerations. This chapter introduces some of the decision alternatives as well as some of the basic building blocks of data representation, and we will treat these topics in greater depth in Part II of the book. The fundamental concept here is that the transition between real world events and entities to digital format requires choices and data management.

For most analytic purposes, we organize data into a two-dimensional tabular format in which each row is an individual observation or case, and each column contains a measurement or attribute pertaining to that individual. There are times, though, when this is not the only option or even the preferred option. Hence, decisions about the structure and content of a data table are driven by the information needs of a particular investigation as well as the intended analytic purposes.

Some investigators are in a position to develop, gather, and structure data from scratch, while others might avail themselves of digital data that was gathered and stored by others. For some studies, the necessary data might be sitting neatly in a single accessible table while for other studies the investigator must combine and cull data from multiple sources. The tasks begin with an understanding of your investigative goals.

Matching Form to Function

As an organizing example (to be revisited in this and later chapters) we will consider data related to members of the U.S. House of Representatives. As a starting point, visit http://www.house.gov/representatives/ for a list of the 435 individuals currently serving in the House. Figure 2.1 shows a portion of a directory in June 2016 (U.S. House 2016).

Figure 2.1 Directory of Members of the U.S. House of Representatives

For each state the website identifies the representative for each congressional district, listing party affiliation, office location, phone number and committee assignments. Within the list, the names are hyperlinked to individual web pages for each member, and these in turn contain varying information about pending legislation, voting records, news articles, videos of speeches and appearances, and links for constituent services and social media among other items. Some of the information might be standard for all members (for example, office locations and contact information), representatives customize their sites as they see fit.

The variety of data and quantity of data behind this directory page are rich, and the table layouts vary as we explore the links. Some of the data is relatively stable over time and some can change frequently. Some changes occur on a regular schedule, such as those related to election outcomes. Other dynamic information happens when it happens, such as news events.

The designers of the directory page in Figure 2.1 made choices about which data elements to display, in what sequence, what to abbreviate and what to spell out. Their purpose was presumably to allow for flexible access to individual members of the House, but a prospective analyst or modeler might want all or part of the related data for another well- or loosely defined purpose. A static or interactive report will have one set of data requirements and challenges, while a statistical modeling project will likely call for different data requirements. Data mining projects or business intelligence applications have yet different goals and therefore different data needs.

In this chapter we'll introduce some basic ways of classifying and organizing data. Later chapters revisit these options in greater depth and provide relevant JMP instructions.

JMP Data Tables

In the JMP environment the fundamental framework for data organization is the *data table*. A JMP data table is a two dimensional grid in which each column is typically a variable and each row an observation or case. JMP recognizes that data elements have unique properties and can play particular roles within an analysis. Some distinctions are rooted in considerations of efficient data storage, and others constrain or define the set of analytic tools that are applicable. For example, we compute the mean and standard deviation of continuous numeric data, but not for discrete nominal data.

The different data classifications are expressed as attributes or properties of a column, and the two preeminent properties are the *Data Type* and the *Modeling Type*. When a data table is created or when a new column is added to a data table, you must specify these properties within the Column Info window (Figure 2.2).

Figure 2.2 The Column Info Window

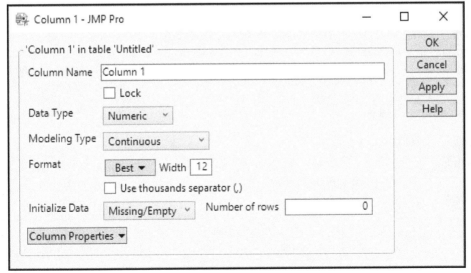

Data Types and Modeling Types

Users have the option to select and modify numerous attributes for any column. But at the very least, they must specify an initial data type and modeling type for each column. When data are read into JMP from another format, or pasted or typed, JMP makes an intelligent guess about the data type and modeling type. Users are well advised to confirm that the assigned types are consistent with the aims of the analysis.

> Veteran JMP users take note. JMP 12 introduced the "Expression" data type, and JMP 13 adds several new Modeling Types, described here.

Data Types

A column's data type affects how values are displayed within the data grid display, how much space the values consume internally (on disk or in memory), and whether the values can be used in calculations. Data types might refer to raw data or to the result of computational or logical formulas. Users select from these four data types:

- **Numeric**: Values are intrinsically quantitative data, suitable for computation.
- **Character**: Values are alphanumeric and might be entirely numeric but still not suitable for computation. For example, a telephone number might consist entirely of numeric digits, but still not be a number.
- **Row State:** Values indicate, for example, whether rows are excluded, hidden, and so on. They are not raw data values, but, rather, they refer to user-specified options.
- **Expression**: Entries are JMP Scripting Language (JSL) code chunks, which allow a large range of possibilities. The underlying code might represent a histogram, a picture, a matrix, or various other objects.

Modeling Types

In addition to the data type, each column in a JMP data table has a pre-specified modeling type that governs how the column is treated in analyses. Table 2.1 lists and describes the available modeling types. It provides some generic examples as well as illustrative examples based on the House of Representatives directory. Note that some of the examples in the right-most column draw on data typically found at the websites of individual representatives or other databases maintained by the U.S. Congress.

Table 2.1: JMP Modeling Types

Modeling Type	Data type(s) and Description	Typical Examples	House of Representatives Directory Example
Continuous	Numeric data only. Used in operations like sums and means.	Height Temperature Time	Year first elected; age; longitude and latitude of home district office
Ordinal	Numeric or character data. Values belong to ordered categories. For numeric data, default order is numerical; for character data, default order is	Month (1,2,...,12) Letter grade (A, B,...F) Size (small, medium, large)	Bill number of pending legislation

Modeling Type	Data type(s) and Description	Typical Examples	House of Representatives Directory Example
	alphabetical. User can specify alternative ordering.		
Nominal	Numeric or character data. Values belong to categories, but the order is not important.	Gender (M or F) Color Test result (pass or fail)	District number; Room number; Party; office building (CHOB, LHOB, and so on)
Multiple Response	Character data. Values must be separated by commas (or a user-specified character)	Degrees earned, job titles held	Committee assignments
Unstructured Text	Character data. Unique text, phrases or longer bodies of prose, that are not suitable for categorical analysis.	Item description Tweets User feedback or recommendations	Title of pending legislation; press release text
Vector	Expression data only. Entries are column or row vectors		
None	Any data type. Used when column entries are not well represented by other modeling types.	Photograph ID values	Photograph; email address

Within the column pane of the data window, each modeling type has a unique icon. For an example within the Sample Data directory, open the **Big Class Family** data table. Figure 2.3 shows the Columns pane illustrating the icons.

Figure 2.3 Modeling Type Icons

Long-time JMP users are familiar with the blue triangle for continuous data, the red bar chart for nominal, and the green ascending bars for ordinal. Newer symbols include a white square for **None** modeling type (next to **Picture**), the red square and X for **Multiple Response** (**sibling ages**, **sports**, **countries visited**, and **family cars**), the black dashes for **Unstructured Texts** (**reported**

illnesses) and square brackets for **Vector (age vector)**. In this example, the **age vector** results from a formula that creates a 1 x 6 vector of 0s and 1s, representing ages 12 through 17.

For continuous numeric data, JMP also provides formatting options, including numerical, financial, date, time, duration, and geographical. For character data, you can alter value labels, specify an ordering sequence, or assign colors to particular levels of a categorical variable.

If we wanted to create a JMP data table that draws on some of the data available through the House of Representatives portal, we'd need to navigate multiple pages. Each page would need to draw on one or more databases containing information about House members. The navigational task is easier with some rudimentary understanding of the most common database structure: the *relational database*.

Basics of Relational Databases

The relational database has been the workhorse of data storage for decades (Codd 1990). The relational model divides data elements into related two-dimensional tables linked together with shared variables known as *keys*. Although the advent of big data has challenged the capacities of the standard relational model, much of the data that an analyst is likely to access are still in relational databases (Asay 2016). It is beyond the scope of this chapter to provide a full treatment of the relational model. But it might be helpful to understand some fundamental concepts as a context for the processes of acquiring and preparing data for analysis and modeling. Readers with a background in database management should skip this section. Chapter 5 is devoted to methods of accessing data directly from databases, and JMP offers an extensive database querying functionality. Anytime we want to combine data residing in separate tables, a working understanding of the relational structure is valuable.

A good starting point is the *entity-relationship* concept, introduced by Chen in 1976. We can think of a data model as containing information about related entities, which are subjects or objects in the domain of interest. For example, in a hospital setting, entities might include patients, doctors, nurses, rooms, equipment, diagnoses, procedures, medications, and the like. Each entity will have a series of *attributes* or relatively stable characteristics of interest, such as (for people) name or year of birth. At least one attribute must uniquely identify an entity.

Entities have relationships with one another: one doctor might have many patients, and a patient might have several doctors. Similarly, various patients might receive the same treatment, and any patient might receive many treatments. Such relationships are known as "many-to-many" relationships. Some relationships are "one-to-one": At any time, only one patient can be assigned to a single hospital bed. Yet other relationships can be "many-to-one."

In the relational model, each entity warrants its own table of data. So, for example, there might be one table in the database for basic patient data. Essentially, a database design ordinarily conforms to the *third normal form* (Codd 1990) of relational databases:

1. Each variable forms a column.
2. Each observation forms a row.
3. Each type of observational unit (entity) forms a table.

One column in an entity table must be a *primary key* field that uniquely identifies the instance or observation of the entity. For example, every patient would have a unique patient number and every medication a unique code. Other columns might optionally include *secondary keys*, which link one table to another.

Consider this example: A patient table can contain each patient's full address, including (in the U.S.) street, city, state, and ZIP code. Alternatively, and potentially with added storage efficiency, the patient table could hold the street and ZIP code, where the ZIP code functions as a secondary key linked to another table of ZIP codes, cities and states. In this way, the system can use the postal code to determine the city and state, and avoid repetitive storage of those data elements within the database.

In addition to the tables representing entities, there will be other tables representing transactions or other relationships. For example, there could be a table that tracks each medication dispensed to a given patient. The table might have columns representing the date, time, patient code, medication code, quantity dispensed, prescribing doctor code, and dispensing nurse code. Notice that most of the columns would be secondary keys pointing to other entity tables containing further relevant attributes.

Returning to the House of Representatives data, entities would include the House members, states, committees, political parties, bills (proposed legislation), office buildings, speeches, events, and more. The main directory site shown in Figure 2.1 illustrates some of the attributes of an individual representative, such as name and telephone number. Some of the columns are not attributes as such, but rather point to other entities: party affiliation, room number, and committee assignments. Note also that the entries in the Name column on the website link to members' individual websites, which in turn are points of entry to huges amounts of related data.

The details of this specific example are less important than the general concept. With a working understanding of the relational model, you can begin to picture the structure of the data that underlies the simple table in Figure 2.1.

For most analyses in JMP, the basic tabular format with variables in columns and observations in rows is the standard approach. There are exceptions, and these are treated in detail in Chapter 7. Some JMP analysis platforms handle more than one value per table cell, such as those for multiple response modeling types. As we move through the next several chapters, the concepts of linked data tables, third normal form, data types, and modeling types will recur.

Conclusion

This brief chapter has introduced some of the important concepts that form the basis of the chapters that follow. It has discussed the usefulness of storing data in consistent tabular formats, and the vital relationship between the format of a data table and the analytic purpose at hand. Additionally, we've covered the key elements of data types and modeling types. Finally, we've provided the basic foundations of the relational database model, a structure that paves the way for accessing data and for combining data from numerous sources. All of these ideas come into play in the chapters that follow, starting with the next chapter about sources of data.

References

Asay, M. 2016. "NoSQL keeps rising, but relational databases still dominate big data." *TechRepublic.com*. Available at http://www.techrepublic.com/article/nosql-keeps-rising-but-relational-databases-still-dominate-big-data/.

Chen, Peter Pin-Shan. 1976. "The Entity-Relationship Model—Toward a Unified View of Data." *ACM Transactions on Database Systems* **1** (1): 9-36.

Codd, E.F. 1990. *The Relational Model for Database Management: Version 2.* Reading, MA: Addison-Wesley.

SAS Institute Inc. 2016. "Understanding Modeling Types." Available at http://www.jmp.com/support/help/Understanding_Modeling_Types.shtml.

U.S. House of Representatives. 2016. "Directory of Representatives." Available at http://www.house.gov/representatives/.

Chapter 3: Sources of Data and Their Challenges

Introduction

Many analytic projects rely on data from disparate sources, or at least involve drawing upon existing data sets for a new purpose. Data come from different sources. Transferring data into JMP is often straightforward, but each type of source can present its own hurdles.

When data are already in saved JMP data tables, there might still be data management hurdles facing the analyst who wants to combine multiple tables or who needs to clean or otherwise pre-process the data prior to modeling. This chapter focuses on the challenges of working data sources other than JMP, and we'll return to JMP tables later in the book.

This very brief chapter surveys common data sources and some of the technical, organizational, and ethical issues connected to the use of data from various sources. Part II of this book explains in detail how you can use JMP to import and start to prepare data from the types of originating sources described here. The goal in these next few pages is to orient the reader to the range of options available, and to provide a preliminary look at the type of planning required to proceed with a project.

Internal Data in Flat Files

Most users and organizations have useful data already residing on their own networks. There are two main types of technical challenges related to pulling data into JMP. First, the data ideally should be organized so that variables occupy separate columns and individual observations occupy separate rows. Second, because different programs store data in distinct and often proprietary formats, the import process must be capable of translating various formats and preserving metadata for use in JMP.

Even when an organization has possession of data, individuals within the organization might have differential authorization to access, view, or analyze the data. Within some organizations, there can be political or institutional barriers to data sharing, and these can make technical barriers look trivial.

JMP cannot help with the latter obstacles, but does offer methods for accommodating the first two types. Chapter 4 walks you through several examples of working with data in different formats, and Chapter 7 extensively covers reshaping and rearrangement of data within a table.

Relational Databases

In a relational database, the variables needed for a given study or analytic purpose will typically reside across several linked tables. Before building models, you first need to select one or more subsets of the available data via database queries.

Though you might create a query table within the native database management system, you can also do so within JMP. Versions of JMP since JMP 12 provide considerable functionality for querying databases of different types. Query Builder is the key tool here, and it is described at length in both Chapters 5 and 8.

External Data on the World Wide Web

For many projects, you might want to incorporate data found on the Internet, whether from regularly published reports or from automatically generated and continuously refreshed sensors. In recent years, the availability of such data has grown rapidly and affords analysts new opportunities for modeling and knowledge discovery through data analytics.

For better or worse, the standards for storage and dissemination of data via the World Wide Web are still evolving, as is JMP functionality in recognizing and accessing data from websites. The ease of obtaining usable web data and converting it to JMP format depends on the design and structure of the particular website.

User-Facing Query Interfaces

Particularly with governmental or public domain websites, users can query the site to generate a table data to download. Some sites provide multiple format options, such as Excel, comma-separated (CSV), SAS, SPSS, and so on.

In later chapters, we'll use some variables extracted from the World Development Indicators, a massive and extensive collection assembled and published by the World Bank (World Bank 2016). You can download a compressed version of the entire data set, including a data dictionary and technical notes, either in Excel or CSV formats. Alternatively, you can query the Indicators database, and download only the portions relevant to your study. For example, suppose your interest is climate change, and you want data about the level of nitrous oxide emissions from each nation over time.

Use your browser and navigate to http://data.worldbank.org/data-catalog/world-development-indicators. After familiarizing yourself with the page, look in the upper right and click the variable selection icon, as shown in Figure 3.1.

Figure 3.1: Access Options

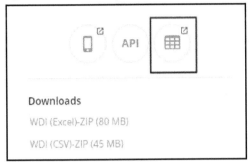

This takes you to a user-friendly, multilingual query page, shown in Figure 3.2. On the left, the user selects rows and columns, and on the upper right the user can choose a download format for the data. With this type of query, much of the initial data management can be handled right at the source. Some of the most crucial choices are listed below.

Figure 3.2: The World DataBank Query Page

1. Select your language from among five options.
2. There are three main options for formatting the downloaded file: Excel, CSV, or tab-separated text. Readers should explore additional options about what to download. The drop-down list of options becomes activated after you specify the content and layout of the data table.
3. Much of the work happens under the **Variables** tab. (See items 5 through 7.) At this point, you select the countries to include.
4. Under **Layout**, you determine the organization of the query result by deciding what goes into columns, rows, and pages. This tab offers several other formatting options.

By default, the tool will create separate columns for each year, rows for series (variables), and pages for countries. This is entirely under user control, and can be customized for any study. In Chapter 7, we see how to reorganize a data table to alter its initial layout for a particular analytic purpose. If you can arrange the data suitably at this stage, that will save effort and resources later. For this example, we'll use the default settings.

5. When you're selecting countries for inclusion, you should be aware that the default list includes 217 nations as well as 47 regional aggregates. (See, for example, "Arab World" in the initial view.) For this example, I'll choose Countries and click the check box icon to select all countries.

6. **Series** refers to the available time series; I've searched on the keyword "nitrous" and selected three series that report emissions in thousands of metric tons of CO_2 equivalent. (See Figure 3.3.) Note also that the information icon next to each series title opens a pop-up window with detailed metadata about the variable.

Figure 3.3: Detail of a Series Selection

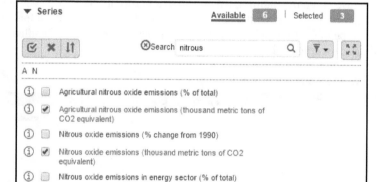

7. **Time** enables you to specify a range of years for inclusion. The data series began in 1960. For this example, I've chosen the most recent 25 years.

    a. In the large Preview area on the right side of the screen, click **APPLY CHANGES** to generate a preview of the table. Finally, click **Download Options** and choose **CSV**.

    b. This downloads a compressed (zipped) file generically named "Data_Extract_From_World_Development_Indicators," which contains two files: one with the raw data, and the other with metadata about the selected series.

    c. Open the downloaded file and extract (unzip) the contents to a directory of your choice.

    d. Finally, within JMP, select **File ▶ Open** and navigate to the directory containing the newly unzipped **Data_Extract_From_World_Development_Indicators** folder.

    e. Choose the larger of the two files and open it.

The newly opened JMP data table contains 29 columns and 656 rows. The first four columns identify the country and series by name and identifier code, and the remaining 25 columns are the annual measures. In Chapter 7, we'll see how to rearrange the data if so desired.

Tabular Data Pages

Some websites lay out data in tables—everything from recent stock prices to census data to the standings of sports teams or barometric pressure measured hourly at a particular location. It is easy enough to locate such pages. With the JMP **Internet Open** command, it can be a very straightforward task to convert the data on the web page into a JMP data table. Chapter 6 works through several relevant examples.

Evolving WWW Data Standards

Readers should understand that the standards for presenting and exchanging data on the World Wide Web continue to evolve. This can sometimes present obstacles to obtaining desired data sets, though there are reasons to be optimistic that these obstacles will become fewer in number as time goes by.

Here is view of the World Wide Web Consortion (W3C, 2016):

> In addition to the classic "Web of documents" W3C is helping to build a technology stack to support a "Web of data," the sort of data you find in databases. The ultimate goal of the Web of data is to enable computers to do more useful work and to develop systems that can support trusted interactions over the network. The term "Semantic Web" refers to W3C's vision of the Web of linked data. Semantic Web technologies enable people to create data stores on the Web, build vocabularies, and write rules for handling data. Linked data are empowered by technologies such as RDF, SPARQL, OWL, and SKOS.

In short, this is a period of significant change. The availability of granular data on a huge variety of topics continues to explode, and the fundamentals of data storage and access likewise continue to grow and change. The methods described in this and later chapters will no doubt be surpassed by faster, more flexible, and more efficient methods in due time.

Ethical and Legal Considerations

Before stitching together data drawn from multiple sources, analysts should give careful consideration and seek authoritative advice about the risks of ethical and legal pitfalls. This book cannot provide legal advice, but there are several general guidelines that apply, particularly when using data for purposes that were not imagined when the data were originally gathered. Among these guidelines are the following:

- Just because a data set is on the web doesn't necessarily mean it can be used for any and all purposes.
  - For example, Hildebrandt asks "Does the fact that personal data are publicly available render their exploitation and monetization ethical, or does it?" (See Hildebrandt 2013, p.8.)
- Similarly, availability on the web does not guarantee that data are current, complete, accurate, or reliable.
  - boyd and Crawford (2012) cite the example of Twitter feeds that appear to be complete but might not be.
- Before using data from any source, the analyst is responsible for discovering how, when, where it was assembled or captured, and by whom.
- Learn the meanings of the terms "public domain," "fair use," and "commercial use."

In the realm of ethics, one good place to start is the "Ethical Guidelines for Statistical Practice" promulgated in 1999 by the American Statistical Association. Practitioners who are unfamiliar with this document would do well to review it. The ASA Guidelines cite eight general areas including the following:

- Integrity of data and methods, which includes proactively ensuring the reliability and authenticity of data in any study.
- Responsibilities for accuracy, transparency, diligence, and judgment as work relates to the following:
  - Funders, Clients, and Employers
  - Publications
  - Research Subjects

Data security and privacy are two of the major areas of ethical concern, but there are others as well. There is considerable research on the ethics of analytics in the era of big data, particularly surrounding clickstream and other automatically or passively generated data. The potential for stumbling into ethical gray zones when combining data from multiple sources for innovative purposes. For example, one common way to protect individual privacy is the use of informed consent: Fully and clearly explain to an individual precisely what data are being captured and for what purpose. Yet how can an individual consent to uses that have not yet been conceived, or which involve methods that are not easily explained in simple, plain language?

The issues related to ethical uses of data are numerous and complex. Readers are encouraged to investigate some of the publications referenced at the end of this chapter, particularly those by Hildebrandt and by Nissenbaum.

Conclusion

This chapter has reviewed some of the challenges corresponding to the most common sources of data. Before this book dives more deeply into the techniques of obtaining and wrangling data, the chapter has sketched some of the issues to be considered later in greater detail. The chapter serves as both a guide and a cautionary note about the tasks ahead.

References

American Statistical Association. 2016. "Ethical Guidelines for Statistical Practice." Retrieved from http://www.amstat.org/about/ethicalguidelines.cfm.

boyd, danah, and Kate Crawford. 2012. "Critical questions for big data: Provocations for a cultural, technological, and scholarly phenomenon." *Information, Communication & Society* 15:5, 662-679.

Hildebrandt, Mireille. .2013. "Slaves to Big Data. Or Are We?" *The Selected Works of Mireille Hildebrandt.* Available at https://works.bepress.com/mireille_hildebrandt/52.

Munzert, Simon, Christian Rubba, Peter Meissner, and Dominic Nyhuis. 2015. *Automated Data Collection with R: A Practical Guide to Web Scraping and Text Mining.* Chichester, West Sussex, UK: John Wiley.

Nissenbaum, Helen. 2009. *Privacy in Context: Technology, Policy, and the Integrity of Social Life.* Stanford, CA: Stanford University Press.

World Bank. 2016. "World Development Indicators." Available at http://data.worldbank.org/data-catalog/world-development-indicators/.

World Wide Web Consortium. 2016. *Standards.* Retrieved from http://www.w3.org/standards.

Chapter 4: Single Files

Introduction

In this chapter and in the two chapters that follow, we'll see how to transfer data from various sources into a JMP data table. This chapter focuses on common data file formats that analysts are likely to encounter when assembling data for a project. Chapter 5 covers database queries, and Chapter 6 demonstrates methods for reading data from websites.

In this chapter, we assume that the starting point is a file residing in a local directory, and we consider different methods for transferring the data from one format into a $*$.jmp data table file. Another implicit assumption is that projects often call upon the analyst to assemble and combine data from more than one source file, often in collaboration with other analysts. Consequently, we'll treat each example as if it is part of a larger effort to assemble a set of $*$.jmp files for further analysis and to document steps taken.

JMP is able to read a wide variety of file formats compatible with popular statistical and database packages. This chapter will demonstrate the importation of three common formats for JMP users—namely Excel, text, and SAS data files.

Review of JMP File Types

In JMP, we store raw data in tables that contain the data values as well as metadata about the structure and content of the file. This chapter deals exclusively with data files, but users should also be aware of several other file types within the JMP environment to store scripts, journals, reports, and SQL queries. As illustrated in Figure 4.1, the JMP **File > Open** command allows for numerous file formats. You can readily open a text file for analysis. But it is much more efficient to open the file; specify data and modeling types; adjust formatting; and then save (re-write) the file and its metadata as a $*$.jmp file.

Figure 4.1: File Formats Readable in JMP

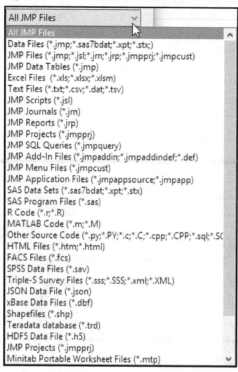

The list in Figure 4.1 includes data files, databases, and scripts (program files). But, in this image, the list is truncated. Below the Minitab Portable Worksheet Files, JMP also lists dBase and MS Access database files. The list is quite extensive and has grown with successive releases of JMP. Imports are more or less straightforward. In this chapter, we'll concentrate on Excel, text, and SAS files. Data originally stored by other statistical software applications such as SPSS and Minitab can also be imported, or alternatively can be saved in Excel or text formats for transfer. At the end of the chapter, we'll briefly review some of the ways that you can save a JMP data table for use with other software.

In the next three sections, we'll sequentially demonstrate methods to convert a single data file into a JMP data table. Each section provides examples using illustrative data, but readers are encouraged to apply the methods to their own data files. No single example can illustrate all of the idiosyncrasies of real data, but the illustrations aim to highlight common problems and issues. Ultimately, readers might need to refer back to relevant JMP documentation and be prepared to explore and experiment.

One recurring theme in the next few sections is *reproducibility* (Peng 2012). It is not enough to successfully convert a data file into a JMP data table, particularly when you are working on a team or can anticipate a need to convert the same or a similar file at a later time. Documentation is vital: How was the conversion accomplished? What customizations were needed? What adjustments were made to modeling types or other characteristics of the data? Within each example, we also find directions for preserving a record of the conversion to that it can be replicated, with modification if necessary, by someone else at another time.

Common Formats Other than JMP

MS Excel

Perhaps the most common format for storage of simple tabular data is that of Microsoft Excel, widely used throughout the world. The installed user base enormous, and many websites offer downloads of data in *.xlx and *.xlsx formats. As a consequence, many analytics projects include data that reside in an Excel spreadsheet. There are two main strategies for moving data from an Excel sheet into a JMP data table. In the first, we start from a JMP session and import the data. In the second, available only in the Windows environment, you can use the JMP Excel Add-In to transfer all or part of a worksheet from Excel to JMP.

Importing an Excel File

In the simplest case, an Excel file is already in tidy format, with each column containing a single variable and each row an observation. In fact, the JMP default settings look for column headings in the first row of an Excel file with data starting in the second row. Because you cannot always depend on finding such a clear layout, our first example will illustrate the process with a spreadsheet that is slightly untidy.

Presumably readers have their own Excel files aplenty to use. But in this chapter, we'll use an Excel file downloaded from the United Nations Population Division, Policy Section. The downloaded file is in Excel format, providing information for nearly 200 countries of the world as of 2013. The UN revises the table every two years, drawing upon the biennial revisions from the World Population Database, on a variety of topics from population size, age distributions, public health, migration, and the like. In total, there are 47 policy variables for each country, downloaded as an Excel format file. We use this to illustrate what might be found in an internal corporate or agency directory, as well as JMP facilities with importing Excel format.

The original Excel file is available for download at https://esa.un.org/PopPolicy/wpp_datasets.aspx, and Figure 4.2 shows a portion of the spreadsheet in Excel. The Excel workbook contains a single tab, titled **rptWebDataQuery**.

Figure 4.2: 2013 Revision of the World Population Policies Database Extract

	A	B	D	E	F	G	H	I
1	World Population Policies Database: All Variables				2013 Revision			
2	Country name	Country code	Region	Development level	Least developed country	Policy on growth	Level of concern about ageing of the population	Measures adopted to address population ageing
3						2013	2013	2013
5	Afghanistan	4	Asia	Less dev. region	Yes	Lower	Minor concern	Neither
6	Albania	8	Europe	More dev. region	No	Maintain	Major concern	1,2
7	Algeria	12	Africa	Less dev. region	No	Lower	Minor concern	Neither
8	Andorra	20	Europe	More dev. region	No	Maintain	Minor concern	2
9	Angola	24	Africa	Less dev. region	Yes	Maintain	Minor concern	No data available
10	Antigua and Barbuda	28	Latin Amer. & Caribbean	Less dev. region	No	No intervention	Major concern	2
11	Argentina	32	Latin Amer. & Caribbean	Less dev. region	No	No intervention	Minor concern	2
12	Armenia	51	Asia	Less dev. region	No	Raise	Major concern	2
13	Australia	36	Oceania	More dev. region	No	No intervention	Major concern	2
14	Austria	40	Europe	More dev. region	No	Raise	Major concern	1
15	Azerbaijan	31	Asia	Less dev. region	No	Raise	Major concern	1,2
16	Bahamas	44	Latin Amer. & Caribbean	Less dev. region	No	Maintain	Major concern	2
17	Bahrain	48	Asia	Less dev. region	No	Lower	Minor concern	Neither
18	Bangladesh	50	Asia	Less dev. region	Yes	Lower	Major concern	Neither
19	Barbados	52	Latin Amer. & Caribbean	Less dev. region	No	No intervention	Major concern	1
20	Belarus	112	Europe	More dev. region	No	Raise	Major concern	2
21	Belgium	56	Europe	More dev. region	No	No intervention	Major concern	1,2
22	Belize	84	Latin Amer. & Caribbean	Less dev. region	No	Lower	Major concern	Neither
23	Benin	204	Africa	Less dev. region	Yes	Lower	Major concern	Neither
24	Bhutan	64	Asia	Less dev. region	Yes	Lower	Minor concern	Neither
25	Bolivia (Plurinational State of)	68	Latin Amer. & Caribbean	Less dev. region	No	No intervention	Major concern	1,2
26	Bosnia and Herzegovina	70	Europe	More dev. region	No	No intervention	Major concern	1,2
27	Botswana	72	Africa	Less dev. region	No	Maintain	Minor concern	Neither
28	Brazil	76	Latin Amer. & Caribbean	Less dev. region	No	No intervention	Major concern	Neither
29	Brunei Darussalam	96	Asia	Less dev. region	No	No intervention	Minor concern	1,2
30	Bulgaria	100	Europe	More dev. region	No	Raise	Major concern	1,2
31	Burkina Faso	854	Africa	Less dev. region	Yes	Lower	Minor concern	Neither
32	Burundi	108	Africa	Less dev. region	Yes	Lower	Minor concern	Neither

Notice at the outset that the data in this spreadsheet begin in row 5, and that the column headings seem to straddle rows 2 through 4. Moreover, though not visible in Figure 4.2, following the last country observation for Zimbabwe in row 201, there are two footnotes occupying cells A204 and A205.

The fact that these tables are updated biennially also means that it will be wise to anticipate the need to repeat the import process periodically in the future. So whatever approach we take to reading in the data should be reproducible—which is to say, well-documented for future reference. You might selectively **Copy** all of the data and **Paste with Row Names** (**Edit** menu), but there is no obvious way to capture and record the steps taken across platforms to accomplish the import.

In this example, we'll demonstrate the JMP intelligent tools for reading in "foreign" data tables (that is, tables in formats other than JMP). We'll assume that the target file is in a directory on your system. In this instance, I have saved it as **2013_WPPDataset_AllVariables**.

1. Select **File ▶ Open** and navigate to the directory containing the Excel file.
   By default, the **Open Data File** dialog lists all files formatted for JMP as well as those readable as JMP files, so you will see the Excel file listed.
   If you don't see the Excel file, then in the lower right corner of the dialog, click the drop-down menu **All JMP Files**, and choose **Excel Files (*.xls, *.xlsx, *.xlsm)**.
2. Click the file name, and it will appear in the File name box.
3. Now click the down arrow on the **Open** button, and select **Use Excel Wizard**. Figure 4.3 shows wizard dialog in its initial state for this example.

Figure 4.3: The Excel Import Wizard

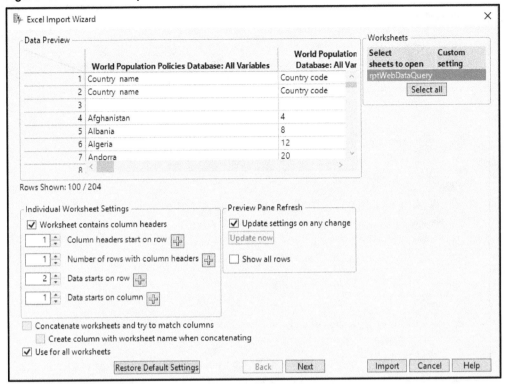

Across the top of the dialog are two regions. The left one is a preview layout of the data as the wizard intends to import it. The right region lists the worksheets within the file. This particular workbook contains a single sheet, but the wizard can handle the importation of one or more sheets from a single workbook, and the user can specify different layouts for each.

Looking at the layout, we see that the titles shown are those found in row 1 of the Excel workbook. The second and third rows in the preview are the actual column headings, and the data about countries don't start until row 4.

The lower portion of the wizard dialog allows us to customize the initial default settings to create an appropriately organized JMP data table. We also note that the preview shows the first 100 rows of data. Having noticed the footnotes at the bottom of the Excel file, let's check the box marked **Show all rows** at this point.

To adjust the layout of column headings and data rows, make the following changes in the **Individual Worksheet Settings** until the dialog looks like Figure 4.4:

1. Column headers start on row **2** of the Excel file.
2. There are **3** rows devoted to column headers
3. The data values first appear in row **5**.

Figure 4.4: Individual Worksheet Settings Updated

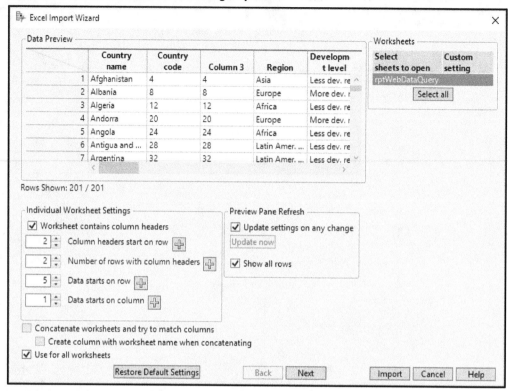

Due to the layout of the headings in the Excel file, the **Data Preview** shows a redundant **Column3**, which repeats the **Country Codes**. We'll want to delete that column later.

1. Click **Next** at the bottom of the wizard.
2. As shown in Figure 4.5, specify that the final data row is row 201 in the Excel sheet, which corresponds to the 197th country Zimbabwe. Then click **Import**.

Figure 4.5: Identifying the Final Data Row for Import

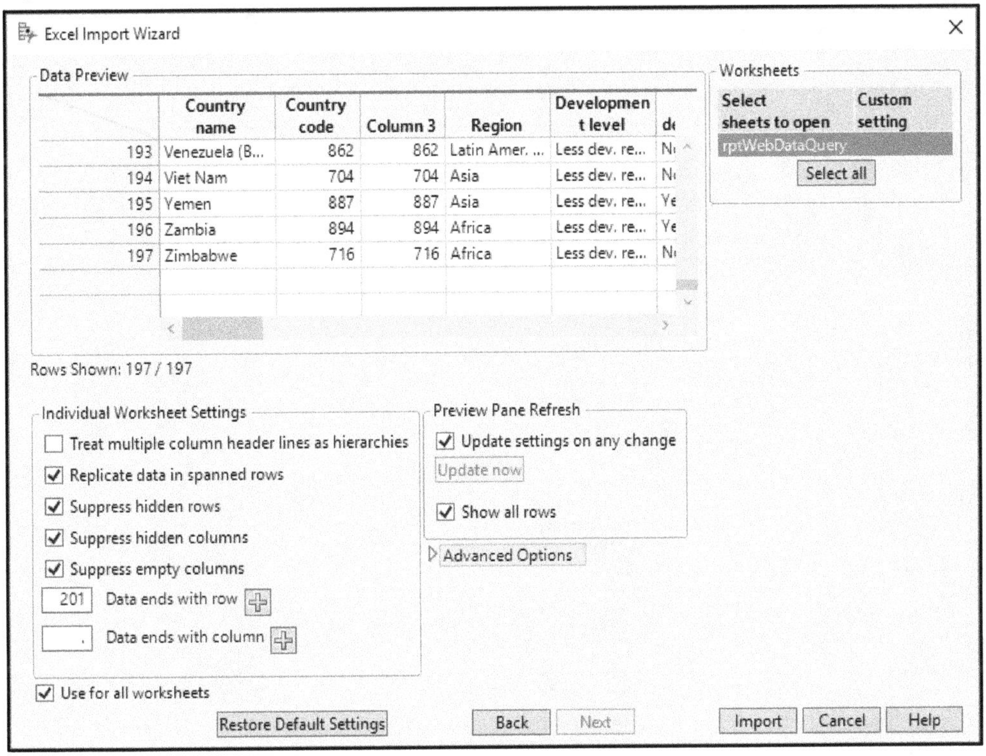

3. Save the results a JMP data table with a suitable name such as **WPP_2013**.

4. Look at the upper left of the data table, shown in Figure 4.6. In the top data table panel, you will see a green triangle and the word **Source**.

   a. Click the green triangle once.

   b. A new data table will open, identical to the one we just imported. That's because JMP saves a copy of the JSL script that was created while we ran the Excel Import Wizard. To see the script, do the following:

   c. Right-click the green triangle and click **Edit**. This opens the script editor shown in Figure 4.6.

Figure 4.6: The JSL Script to Import the UN Data

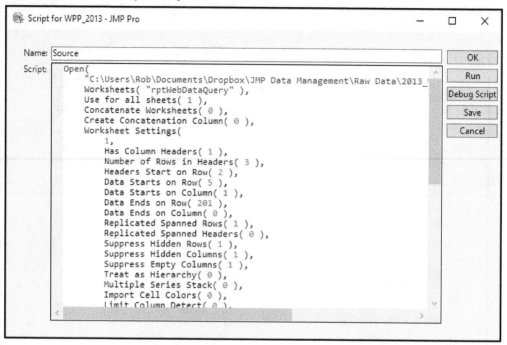

It's not important to read or fully understand each line of code in the script. However, do take note that this is a long **Open** command that identifies an original directory location of the Excel file, and it then lists a series of specifications about the location of data columns and rows. More importantly, JMP preserves a record of the import within the very data table itself. When the UN updates the table for 2015, 2017, and later years, the import instructions will be available to us.

Following the import, it is important to verify data and modeling types as described in Chapter 2, and to confirm that the columns and headings were interpreted consistently with the Excel source file. We already know that **Column 3** in this example duplicates the **Country code** column, so should be eliminated. In this example, you should also note that the ninth column, **Measures adopted to address population ageing-2013**, along with several later columns, contains multiple responses rather than a single value. Depending on the questions to be addressed and the types of analysis conducted, this might warrant further modification using techniques to be discussed in later chapters.

If the Excel spreadsheet has multiple tabs, you can use the wizard to specify the layout of each sheet and to select which sheets to import as separate tables. You can also combine data from multiple sheets into a single JMP data table if there are matching columns—identically named columns common to the sheets involved. Readers are encouraged to read the JMP documentation on "Import Microsoft Excel Files" in the *Using JMP* book within the **Help** menu.

Using the Excel Add-In (Windows Only)

With a spreadsheet open in Excel, you can create a JMP data table and/or launch a range of analyses directly from Excel. This method does not offer the same flexibility as the Excel Wizard, but for a moderately tidy worksheet, it is a quick alternative. Since this chapter deals with ways to take data from your source and create a JMP data table, we'll focus only on the functionality related to data management rather than analysis.

As part of the installation of JMP, Excel users will find a JMP tab on the Excel ribbon, as seen in Figure 4.7. The JMP tab features ten commands, and in this section we'll look at the first two. When used in sequence, these first two commands will create a data table that will open in JMP.

Figure 4.7: The JMP Tab in Excel

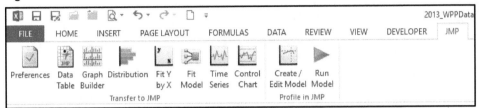

Continuing with the earlier example, we'll now approach the same task. But this time, we begin with the UN data open as the Excel spreadsheet.

1. Within Excel, highlight the area of the worksheet to import. In this example, we will highlight the rectangular area from A2 to AB201.

2. On Excel's JMP tab, click **Preferences**.

Figure 4.8: The JMP Preferences Dialog

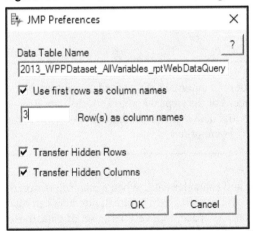

The dialog will suggest a name for the JMP data table, consisting of the concatenation of the Excel file name plus the worksheet tab name. Either leave the default recommendation or change it as you see fit.

3. As illustrated in Figure 4.8, specify that the column names reside in the first three rows of the Excel spreadsheet.

4. Click **OK**.

   The preferences dialog tells JMP what we plan to import. To actually carry out the process, we need the Data Table button.

5. Click **Data Table**.

Now switch to JMP where you will find a new data table like the one shown here in Figure 4.9. Once again, we've transferred 28 columns and 197 rows, and once again we have the redundant third column. This time, country code was misinterpreted as an amount in US dollars. Correct this by changing the data type, modeling type, and format in the **Column Info**.

Figure 4.9: A Data Table Created from Excel

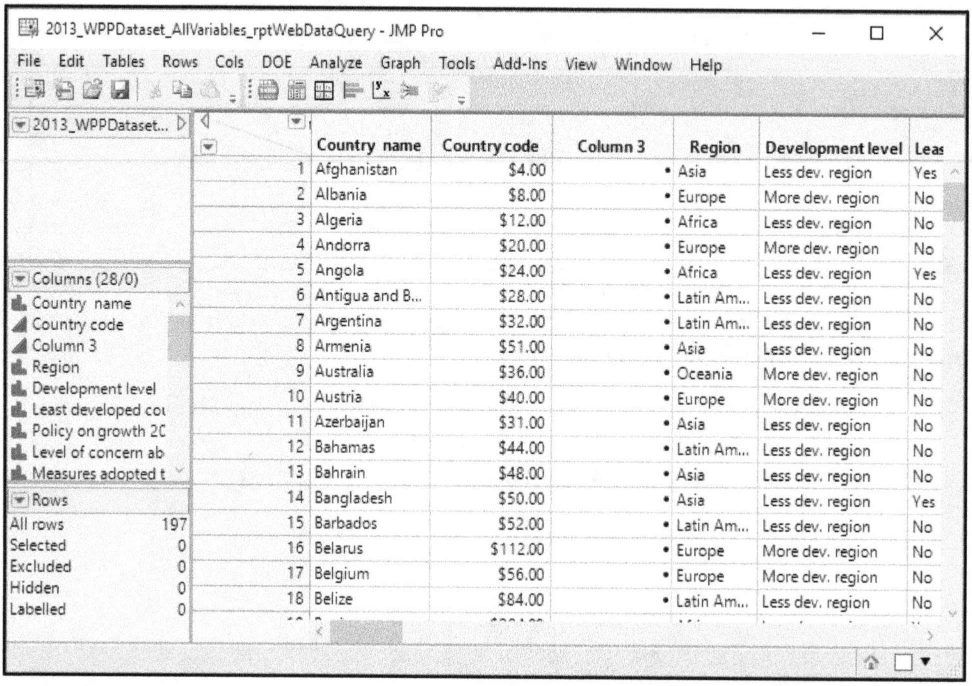

	Country name	Country code	Column 3	Region	Development level	Leas
1	Afghanistan	$4.00		• Asia	Less dev. region	Yes
2	Albania	$8.00		• Europe	More dev. region	No
3	Algeria	$12.00		• Africa	Less dev. region	No
4	Andorra	$20.00		• Europe	More dev. region	No
5	Angola	$24.00		• Africa	Less dev. region	Yes
6	Antigua and B...	$28.00		• Latin Am...	Less dev. region	No
7	Argentina	$32.00		• Latin Am...	Less dev. region	No
8	Armenia	$51.00		• Asia	Less dev. region	No
9	Australia	$36.00		• Oceania	More dev. region	No
10	Austria	$40.00		• Europe	More dev. region	No
11	Azerbaijan	$31.00		• Asia	Less dev. region	No
12	Bahamas	$44.00		• Latin Am...	Less dev. region	No
13	Bahrain	$48.00		• Asia	Less dev. region	No
14	Bangladesh	$50.00		• Asia	Less dev. region	Yes
15	Barbados	$52.00		• Latin Am...	Less dev. region	No
16	Belarus	$112.00		• Europe	More dev. region	No
17	Belgium	$56.00		• Europe	More dev. region	No
18	Belize	$84.00		• Latin Am...	Less dev. region	No

Unlike the wizard, the Excel Add-In does not automatically embed a script into the data table. This means that we don't preserve a record of the steps taken to shift the data from Excel to JMP. The Add-In might be more useful for initial data exploration than for a carefully documented analysis project in which reproducibility is important.

Text Files

Recall the difference between numeric and character data. When a data file is stored on disk, every character in the file requires one byte of storage. In contrast, numbers can often be represented more efficiently, needing far less space. In a very large set of data, those efficiencies can make an important difference in the amount of storage space required. In a JMP (or Excel) file, numeric data are recognized as such and stored compactly, but character data are not.

Within this section, the phrase "text file" refers to files containing ASCII-encoded data in which variables are separated either by fixed spacing (for example, country code is in positions 5-10 of each line) or by using a delimiter like a comma or tab character. The data can be of any data or modeling type, but everything in the file is stored internally as character data. The term text does not indicate prose as it might in a text-mining context; it's simply one way of storing data. You frequently find text files as a convenient transitional format in transporting data from one software platform to another. In general, data in a text file have comparatively little metadata. The file might or might not contain column headings or currency symbols, and as part of the import process, it is up to the user to be aware of the proper formatting and data types.

Somewhat analogous to importing data from Excel, JMP will readily import data in a text file either by using pre-specified preferences, the JMP "best guess" about formatting, or with a **Text Import Wizard**. In this section, we'll briefly review the preferences and the import wizard. JMP recognizes a flat text file by its extension, which in most cases will be `*.txt`, `*.csv` or `*.tsv`.

Setting File Import Preferences

For users who are likely to repeatedly import text files that share layout properties, JMP user preferences settings can be modified for text files. Windows users can set the preferences by selecting **File ▶ Preferences** (keyboard shortcut is **Ctrl-k**) and choosing the **Text Data Files Preference Group**, as shown in Figure 4.10. Mac users get to the same settings by selecting **JMP ▶ Preferences**.

Figure 4.10: Setting Preferences for Text Data Files

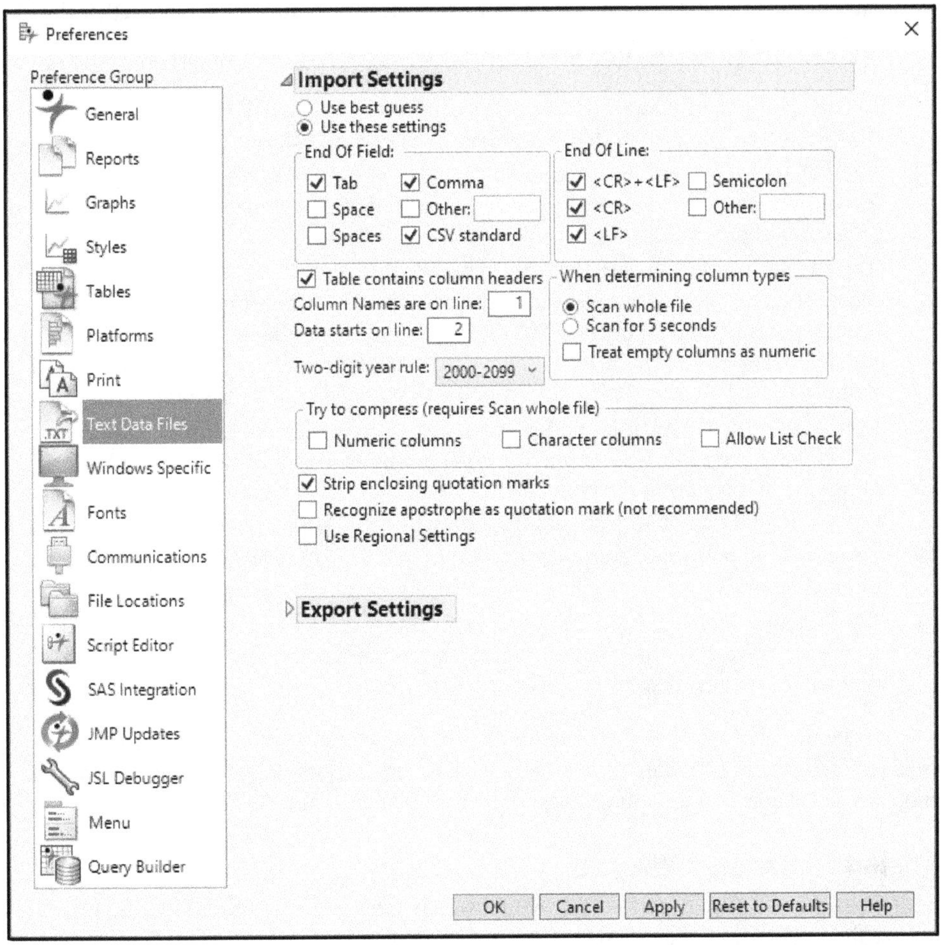

The various option categories in the dialog are summarized as follows:

- **Use best guess:** JMP scans the text file and deploys an algorithm to infer how rows and columns are separated within the file. Users can override the guesses during import.

- **End of Field:** *Field* is a synonym for variable. The user specifies which character or characters act as separators or delimiters between data columns. CSV stands for comma-separated values, a very common way of arranging data in a text file. Commas and tab characters are the two most typical separators. If you open the text file in a standard text editor, commas are visible but the tabs are not.

- **End of Line:** JMP needs to know what signals the end of a row of data. Most commonly, the non-printing *line feed* (LF) and/or *carriage return* (CR) characters are embedded at the end of a data row. If other characters serve this function, the user can so specify.

- **Table contains column names.** As in the Excel wizard, the user can indicate whether and where column headings are located and where the data begin. Notice that this dialog does not allow for nested or multi-line column names.

- **When determining column types:** JMP attempts to infer a data type and modeling type for each column of data. Here the user can indicate how much time the program should devote to this. For files of moderate size, JMP will scan the entire file to inform its guess; for very large files, the user might want to quit after five seconds.

- **Two-digit year rule:** The user specifies how JMP should interpret two-digit years within a date column.

- **Try to compress:** The user indicates whether JMP should store numeric columns in the most efficient way possible. Because the choice of storage format depends on the absolute scale of the numbers, JMP must scan the entire file to determine just how large the numbers are. For character variables that take on relatively few values (for example, month), those values can also be compressed or encoded to save storage space.

- **Strip enclosing quotation marks:** It is often the case that the text file to be imported was created from a different source format. Some software encloses each data value in quotation marks during the conversion to a text format. The user can decide whether to retain the quotation marks or simply drop them.

- **Recognize apostrophe as quotation mark:** Similar to the previous option, though not often applicable.

- **Use Regional Settings:** This refers to differences in US and European conventions. For example, in the US, it is standard to use a period as a decimal point, whereas in other regions the comma is used for that purpose.

For further details, also see *Using JMP* book found in the **Help** menu. An entire chapter is devoted to **Preferences** (Chapter 13 in JMP13). These are adjustable default settings; when opening any particular file, the user has the opportunity to override the defaults.

Using the Text Import Wizard

Regardless of the text import preferences, you also have the option to use a Text Import wizard to customize the importation of a given text file. To illustrate the wizard, we'll use a comma-separated-values file available for download from the World Bank. The "World Development Indicators" data in this particular file will serve as the basis for examples in several upcoming chapters.

The World Development Indicators

The World Bank was established in 1944 to assist with the redevelopment of countries after World War II. It has evolved into a group of five institutions concerned with interconnected missions of economic development. One important goal of its work is the alleviation of poverty worldwide. As part of that mandate, the World Bank annually publishes the World Development Indicators (WDI) gathered from 214 nations. We will return to the WDI data repeatedly in future chapters.

In the public sector as well as in business, policy makers rely on accurate, current data to gauge progress and to evaluate the impact of policy decisions. The WDI data informs policy-making by many agencies globally, and the World Bank's annual data collection and reporting play an important role in the UN's Millennium Development Goals.

1. Visit http://data.worldbank.org/data-catalog/world-development-indicators/. Under Resources, notice that the full data set is available either as a compressed Excel file or a compressed CSV file.
2. Click **WDI (CSV)-ZIP (45 MB)**.

Figure 4.11: World Development Indicators Download Site

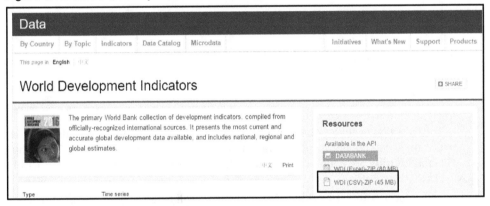

This resource option will download a zipped bundle of seven files, consisting of metadata, footnotes, variable descriptions, and the raw data itself. On a Macintosh, the computer will unzip the files when you click on the file folder. In the Windows environment, it will be necessary to extract the files in the File Explorer. We are interested in a file called "WDI-Data," which comprises indicators reported annually by 214 countries of the world starting in 1960, and since that time some nations have changed names, become independent, or vanished.

As of this writing, there were 1,421 indicators in the file. Though many countries reported only some of the indicators in a given year, the data table contains 214 x 1,421 = 304,094 rows and 60 columns (four to identify countries and indicators, and then one each per year). As we will soon see, many cells are empty, but the data table reserves space for them.

Once you have extracted the folder of seven files, do the following in JMP:

1. Select **File ▶ Open**. Because JMP recognizes `*.csv` files as a JMP data file type, there is no need to adjust the file filter.

2. As shown in Figure 4.12, navigate to the directory where you saved `WDI_data.csv` and highlight the filename.

3. Select **Data with Preview** as the **Open as** option, and then click **OK**.

Figure 4.12: Preparing to Use the Text Import Wizard

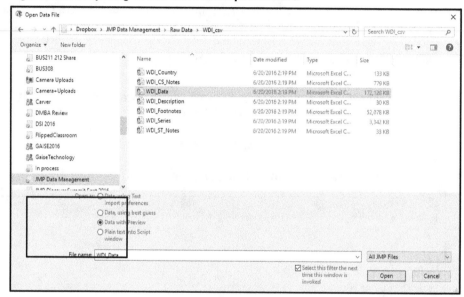

This selection opens a preview window, shown below in Figure 4.13. Initially, we see the first three data columns, and several import options have been suggested. The **Subset** and **Compatibility** areas are hidden by default, but have been disclosed in Figure 4.13 for clarity.

4. In the upper portion (the actual data preview), scroll to the right to see the structure of the data array. Notice also that there are many missing entries and values expressed in scientific notation. Also note that all of the columns are left-justified, which is consistent with their initial character data types.

    JMP knows this is a CSV file, so it expects that the columns are delimited or separated by commas. It does not assume that there are column names, so we should do this:

5. Check the box marked **File contains column names on line** and use the suggested value of **1**.

6. Note that JMP automatically suggests that the data starts on line 2.

    We won't use the **Subset** or **Compatibility** options, but do notice that there are limited choices about how you can import just a subset of the data as well as some of the other items that we saw earlier when setting preferences. This particular file will be a relatively straightforward import. You will surely encounter more complex combinations, but the process is very much the same.

7. Click **Next**.

Figure 4.13: The Preview Window for CSV File Import

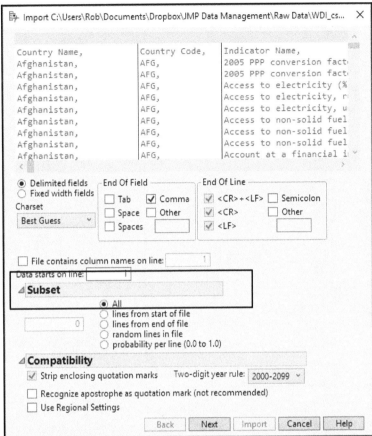

The next dialog (see Figure 4.14) enables us to inspect each column to verify or change the name, data type, and format, or to exclude the column from the import. Depending on the analytic goals, we might choose either an indicator name (verbose), an indicator code (difficult to interpret), or both. In addition, we might want to subset by rows to focus on years of interest.

Because of the arrangement of this file, there is no simple way during importation to selectively import selected indicator data (for example, just the percentage of land area devoted to agriculture or the birth rate). In JMP, we can accomplish this later through use of the Data Filter functionality. Later chapters take up global and local (platform-specific) data filtering.

Take a few moments to click on the data type icons, column names, and (for numeric columns) the red triangles just to see what customizations you can apply. Practice making changes, and then click **Import**. Because this is such a large file, the import will take several seconds.

Figure 4.14: Inspect and Edit Column Names and Types Before Import

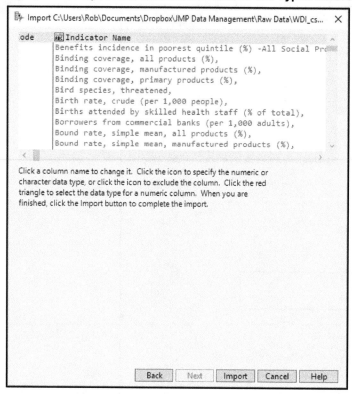

When the import process is complete, you'll see a new JMP data table with 60 columns and 352,160 rows. Notice also that the upper left pane of the data table window again contains a script labeled **Source** (see Figure 4.15). If you open the script in an editor, you will notice that it contains a long **Open** command including the column names, data types, modeling type, and format for every imported column. The script ensures the reproducibility of files imported in this way. As we saw earlier, to inspect the script itself right-click on the word **Source** and choose **Edit**.

Figure 4.15: The Source Script

Following the import, you would **Save** the data table as a *.JMP file with a suitable name for further preprocessing and analysis.

SAS Files

JMP will open files created by several popular statistical packages. In this section, we'll see the general approach by demonstrating the process with a SAS file. Users of SAS are likely to have access to data files created and saved within the SAS environment, and fortunately there are some such files among the sample data supplied with JMP. If you have access to a SAS Metadata Server housing linked files, there is an alternative approach that will be discussed in Chapter 5. In this subsection, we'll restrict attention to a SAS file that is resident on the user's computer or network drive. In the next subsection, we'll see how to import a data table from a server.

Files Stored Locally

1. If you can access SAS files directly from your desktop, select **File ▶ Open** and skip to step 2. If not, continue here. Select **Help ▶ Sample Data Library**. Within the launch window, click **Open the Sample Data Directory**.
    a. Windows: At the top of the dialog, you will see that you are in a directory tree that includes the branches **SAS ▶ JMP ▶ 13 ▶ Samples ▶ Apps**. Go up one level (> **Samples**) and select the folder called **Import Data**. (JMP Pro users will see a JMPPRO folder. Users of earlier JMP versions will see the corresponding version number rather than 13.)
    b. Macintosh: At the top of the dialog, navigate directly to **Macintosh HD**. From there, select **Library ▶ Application Support ▶ JMP ▶ 13 ▶ Samples ▶ Import Data**.
    You will now see a directory containing a large number of files in various formats.
2. In the lower right, click **All JMP Files** and select SAS Data Sets (***.sas7bdat, *.xpt, *.stx**).
    a. Windows: Notice that there is now a choice to use either SAS labels or SAS variable names to define the JMP column names. There is also a check box option to apply table and column properties based on those attributes as defined in SAS 9.4 or later. Use the default options.
    b. Macintosh: The very same options are available after you click the Options button. Use the default options.
3. Find the file named **Diameter.sas7bdat** and click **OK**.

The SAS file simply opens and brings along the relevant metadata. Additionally, JMP includes the import script as a property within the data table as before. Therefore, the process can be repeated at a future time, and the import is fully documented. Also as before, you should save the imported file for further analysis.

Files Stored on a SAS Metadata Server

If the data you need happen to be stored on a SAS Metadata server, the process is similar and straightforward, but requires a couple of preliminary steps. First, you must establish a server connection, and then navigate to the desired file. After that, the import happens much as just described.

This illustration takes advantage of SAS On Demand, connecting to a demo server, and then selecting a table of data related to a customer churn classification task for a fictional mobile telephone service provider. This particular file is used in the SAS Advance Business Analytics course.

1. To establish the connection, select **File ▶ SAS ▶ Server Connections**. This presents the dialog shown in Figure 4.16. Here, I use a **Profile** created for this purpose; SAS users presumably have similar profiles or can create one in the **Manage Profiles** dialog (not shown). Alternatively, you can connect directly with knowledge of the **Machine** name and **Port** number, as shown in the lower portion of the **SAS Server Connections** dialog. Having selected the profile or supplied the machine identification, click **Connect**.

Figure 4.16: Establishing a Connection to a SAS Metadata Server

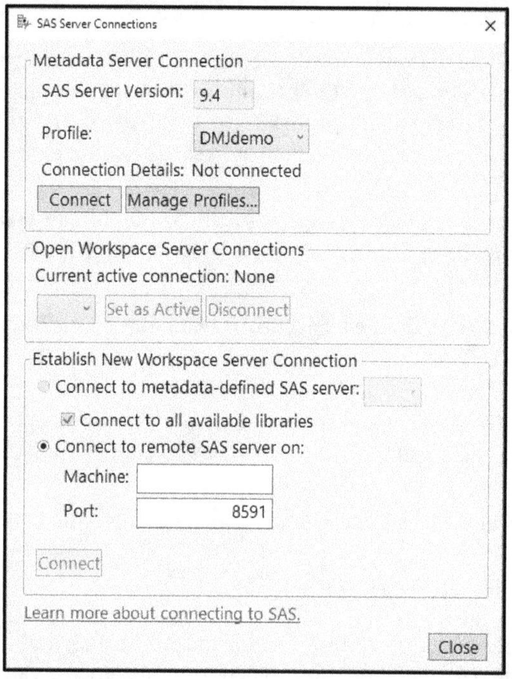

2. Once the connection is established, close the **SAS Server Connections** dialog and then select **File ▶ SAS ▶ Browse SAS Folders**.

   Note: You might equivalently find the data file by selecting **File ▶ SAS ▶ Browse Data**. This platform permits selection of particular columns and rows from a single file. Readers can experiment with both approaches. For the sake of simplicity, this example just uses the former approach because it allows for import of one entire file at a time.

3. As shown in Figure 4.17, we first select the relevant folder and work through the directory tree until we locate the file of interest (**Shared Data ▶ Libraries ▶ ABA1 ▶ Churn** in this case). Then click **Import**.

Figure 4.17: Importing Data Using the Browse SAS Folders Platform

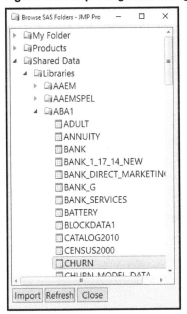

The import might take a few moments, during which time you will see a progress indicator. As with other file imports, JMP uses a best-guess approach for data types and modeling types. In the Table panel of the new JMP data table (Figure 4.18), we find metadata about the original source of the table, with reference to the server, folders path, data set, and the SQL statement, as well as the JSL source code for reproducibility. Chapter 5 has much more to say about SQL and importing data from linked tables.

Figure 4.18: Metadata about a SAS File Imported from a Server

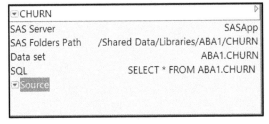

Other Data File Formats

The three file types discussed thus far—Excel, text (csv), and SAS files—are likely to cover a wide range of project needs, but are by no means the extent of JMP data import capabilities. In Figure 4.1, we saw several other common formats. The file import process works analogously to the examples shown above. The particulars of **File ▶ Open** for other file types are documented in detail within **Using JMP** and the **JMP Syntax Reference**, both available among the **Books** in the **Help** menu.

Conclusion

This chapter has explained and demonstrated several typical tasks when you want to import data from a foreign source into JMP for analysis. The process is driven by the analytic goals of the user and the particular format and organization of the original data. As we've seen, the effort and complexity of the process depends largely on the degree to which the originating file is simply organized and contains metadata.

In the next chapter, we take up a slightly more complex importation task: accessing data that reside in a database. Chapter 5 includes a detailed explanation of **Query Builder**, which composes structured queries with a wizard-like interface. We'll mostly use **Query Builder** in the context of a relational database, but it operates analogously with SAS Server files or linked multiple JMP data tables.

References

Peng, Roger D. 2011. "Reproducible Research in Computational Science." *Science* 334 (6060), pp. 1226-1227, 02 Dec 2011.

Truxillo, Catherine. 2012. Advanced Business Analytics, Volume 1Course Notes, Volume 1. Cary NC: SAS Institute Inc.

United Nations Department of Economic and Social Affairs Population Division, Policy Section. 2016. *Population Policies Datasets.* Available at https://esa.un.org/PopPolicy/wpp_datasets.aspx.

World Bank. 2016. "History." Available at http://www.worldbank.org/en/about/history.

World Bank. 2016. "World Development Indicators." Available at http://data.worldbank.org/data-catalog/world-development-indicators/.

Chapter 5: Database Queries

Introduction

This chapter continues in the same vein as Chapter 4. It provides approaches and methods to create a single JMP data table using data extracted from another source. In this chapter, the other source is a relational database. Traditionally, a database user would create a query to specify the data of interest for a particular project. Such a query would generate a new two-dimensional table that would then be read into the available statistics package for further analysis.

If all of the desired data reside in a single table already, the import to JMP is straightforward and analogous to the methods shown in the previous chapter. If the data draw on several database tables, the task is only slightly more complex thanks to the JMP **Query Builder**.

JMP 12 introduced Query Builder to construct and execute multi-table queries directly from a JMP session. Query Builder can expedite work flow by permitting a user to work within JMP to join tables and then select just the data needed for a project while also preserving a complete record of the data transfer.

This chapter walks through some typical methods for moving data from a relational database into a JMP data table, but does not attempt to provide a comprehensive treatment of all features of Query Builder. Interested readers should consult the *Using JMP* book from the JMP Help menu as well as the excellent white paper by Eric Hill (2015), developer of Query Builder. Some of the illustrations in this chapter draw on Hill's paper.

Readers who have access to a relational database in their own work will find the topics in this chapter to be relevant. Rather than following the chapter examples such readers might prefer to run through similar processes using their own data. Readers who lack such access should skim or skip this chapter, which assumes familiarity with database operations as well as availability of access credentials.

Sample Databases in This Chapter

This chapter relies on two illustrative sources, both of which are shared freely by their creators. They are used here to provide context for the techniques demonstrated in the chapter. The particular content is secondary and should help readers perform similar operations on their own databases.

The first database is provided by PostgreSQL, an open-source database (see https://postgresql.org) that is called **dellstore2**. Dellstore2 is a modification of an operational database for a fictitious DVD rental store and was downloaded from http://wiki.postgresql.org/wiki/Sample_Databases. The 15 tables in this database contain information about inventory, customers, stores, films, employees, and vendors as well as rental transactions, film languages, film actors, and so on. Dellstore will serve as our example of a database residing on the same machine or network as JMP.

We also use a data warehouse from Microsoft SQL Server and Azure. Through a freely available Azure account, users gain remote access to several databases including **Adventure Works**, a fictitious bicycle company. The database contains several schemas related to four major scenarios: sales/marketing, product, purchasing/vendors, and marketing. In addition, the Azure environment includes several other demonstration databases. We'll use one containing Olympic medalist data from the years 1900 through 2008.

Since the release of JMP 12 where Query Builder was introduced, the Open Table approach familiar to veteran JMP users has been superseded. For readers with JMP 11 or earlier versions, Open Table is the sole available option and is illustrated in the next section. Other readers are well advised to review "Connecting to a Database" and then skim "Extracting Data from One Table in a Database" as it relates to the Olympic medals example. But then move on to the following section, "Querying a Database from JMP."

Connecting to a Database

There are two ways to initiate a connection between a JMP session and a database, both of which are available by selecting **File ▶ Database**. (See Figure 5.1.) With this selection, you can either open a single table or launch **Query Builder**. Either choice begins by specifying a database connection. The initial connection dialogs are slightly different, and we begin with the simpler and older option, namely **Open Table**. We then demonstrate the process of connecting to a database. In this section, we'll see how to connect both ways.

Figure 5.1: Initiate Database Connection from File Menu

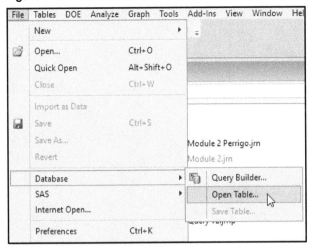

After you click **Open Table**, the **Database Open Table** dialog appears. (See Figure 5.2.)

Figure 5.2: Database Operations for One Table

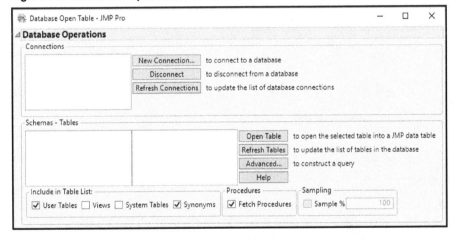

At the top of the dialog, click **New Connection**. This presents the **Select Data Source** dialog (Figure 5.3). This dialog is common to the **Database Open Table** and to **Query Builder**. To make your ODBC connection, you'll need to point to the location of the relevant driver. The ODBC drivers installed on your machine will populate the **Machine Data Source** tab in Figure 5.3.

In this instance, I have one established local data source named PostgreSQL on my computer. If you need to create a connection, you would click **New** in the **Select Data Source** dialog. For Windows users, setting up a new ODBC data source is a standard operation. To set up your data source, select **Control Panel ▶ System and Security ▶ Administrative Tools ▶ Data Sources (ODBC)**. On the Macintosh, you select **Applications ▶ Utilities ▶ ODBC Manager**.

Figure 5.3: Select Data Source

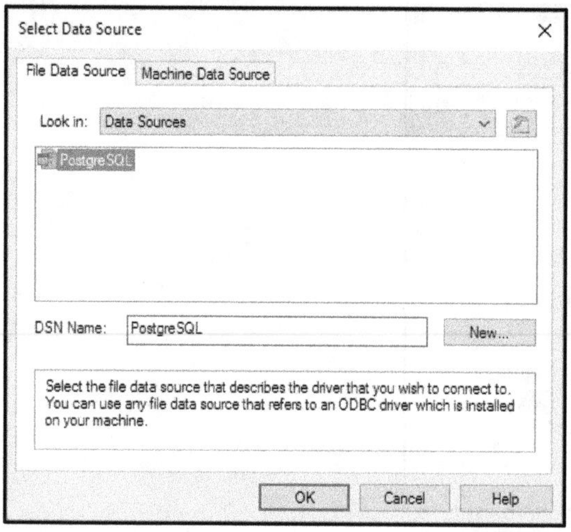

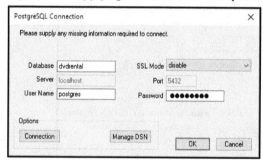

After clicking **OK**, you are prompted to provide user name and password similar to those shown in Figure 5.4. Naturally, these vary by system and user.

Figure 5.4: Supplying Credentials to Complete the Connection

Having provided the necessary information and clicked **OK**, you return to the Database Open Table dialog where you now see the established connection, the available schemas at the connection, and all of the tables available in each schema. (See Figure 5.5.)

Figure 5.5: Database Open Table Dialog Showing a Connection

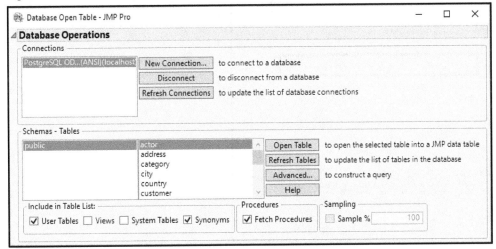

At this point, we'll pause and see how this process differs slightly when using a Machine Data Source.

Readers who want to follow along with the AdventureWorks example should check to see whether the SQL Server Native Client ODBC driver is already present on their machines. If not, it is worth the time to install it.

Continuing with the same initial approach as before, launch the process as shown above in Figures 5.1 and 5.2. In this case, when the **Select Data Source** dialog appears, click the **Machine Data Source** tab as in Figure 5.6. I have previously set up the specifications for a connection to the shared Azure database site, which appears among my machine data sources. Consult your machine settings as necessary, or click **New**. Assuming that readers already have such specifications in place, we'll select the source of interest (**Azure Adventureworks**) and click **OK**.

Figure 5.6: Machine Data Sources

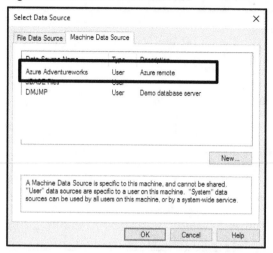

An authentication dialog will appear. (See Figure 5.7.) In this case, it is an **SQL Server Login** asking for a **Login ID** and **Password**. After supplying the credentials, click **OK**.

Figure 5.7: Entering Credentials

The **Database Open Table** dialog (Figure 5.5) will appear again showing the contents of the available schemas and tales.

Extracting Data from One Table in a Database

For some analysis projects, you might find that you only require data from a single table. In such cases, the process is very similar to importing data from a foreign format as illustrated in Chapter 4.

Import an Entire Table

For example, if you want to discover the number of Olympic medals won by each country from 1900 through 2008, you would select the **Medalist** table in the **olympics** schema, found in the Azure connection. (See Figure 5.8.) Highlight the table of interest and click **Open Table**.

Figure 5.8: Selecting a Table to Open

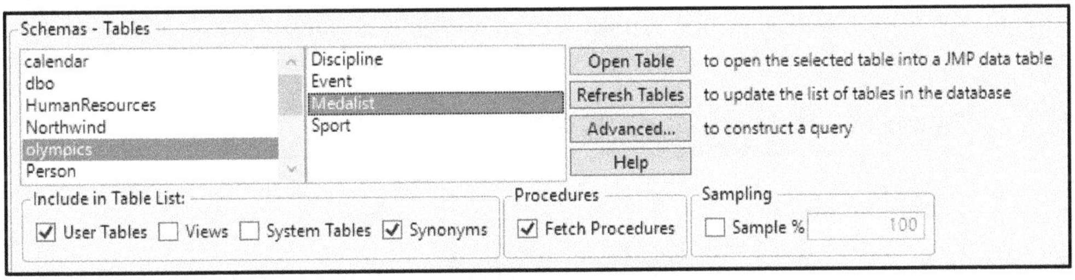

The table will appear in a JMP data table window with familiar layout and features and with a few noteworthy elements. As shown in Figure 5.9, in the left uppermost pane of the window are three green triangles associated with scripts to re-load the table altogether, to update the table from the database, or to save revisions to the database. Once we save the data table locally, these scripts enable us to reproduce the import or carry out updates.

The other noteworthy item here is the identification of the **MedalistID** column as a key column, or **Link ID**, within the database. With a single table, the LinkID property has little impact. If we were to import other tables with a shared Link ID, we could join or perform other database operations.

Figure 5.9: Data Table Imported from a Database

Import a Subset of a Table

In the table we just imported, we note that each row lists one athlete from the Summer and Winter games for all years in the database. Suppose our interest was in Summer Olympic events from 1960 onward, and that we only want to focus on particular columns for analysis. For very large tables, there might be an efficiency gain from importing only the data we actually expect to use rather than importing the entire table.

To choose rows and columns for import, we want to build a query. The **Advanced** button in the Database Open Table dialog (Figure 5.8) provides a tool (**Database Operations** subdialog) to construct and execute a query. In this example (Figure 5.10), the default SQL code will select all rows and columns from the particular table.

1. To choose only some columns, we highlight the columns of interest, and they are added into the SELECT statement within the SQL area of the dialog. Specifically, we want Medalist ID, Event, Edition, NOC, Gender, Medal, and Season.

Figure 5.10: Starting to Build SQL Code

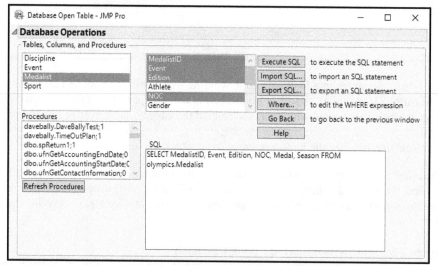

We want Summer Olympic results starting in 1960, so we need a compound WHERE clause filtering by values of two different columns.

2. To choose a subset of rows, we need to add a WHERE clause. Click the **Where...** button, opening the dialog shown in Figure 5.11.

Figure 5.11: Constructing a WHERE Clause

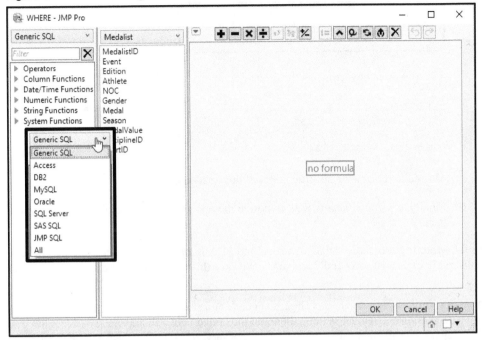

3. In the upper left, notice the drop-down button reading **Generic SQL.** This is a drop-down list of supported vendor SQL dialects. (See inset in Figure 5.11.) Choose the one that corresponds to the data source. For this example, **Generic SQL** is fine.

4. The **Event** column corresponds to year, and we want all years greater than or equal to 1960. Click the gray triangle next to Operators to specify the form of the statement. Choose **A > = B**.

5. In the list of column names, choose **Edition**. At this point, the formula editor area of the dialog should look like Figure 5.12. Move the cursor into the empty gray rectangle in the formula editor, double-click, type 1960, and press **Enter**.

Figure 5.12: Building the Formula in the First WHERE Clause

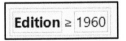

Now we want a second condition, namely to import only rows from Summer Olympic events:

6. In the formula editor, click anywhere outside the formula. In the **Operators** list, click **AND** to add the second part of the statement.

7. Select **A=B** as an operator.

8. From the list of column names, choose **Season**.

9. Finally, move the cursor into the rightmost rectangle, double-click, and type **'Summer'** within single quotation marks. The completed formula will look like Figure 5.13. At this point, click **OK**, which returns you to the **Database Open Table** dialog.

Figure 5.13: The Completed Formula

The SQL area now contains this line of code:

```
SELECT MedalistID, Event, Edition, NOC, Gender, Medal, Season
FROM olympics.Medalist
WHERE olympics.Medalist.Edition >= 1960  AND
olympics.Medalist.Season = 'Summer'
```

This code reflects the specific needs of this analysis.

10. Finally, click the Execute SQL button in the upper right of the dialog box to import the data.

The full **Medalist** data table had 32,591 rows and 11 columns. In this limited import, we have a smaller table of 19,416 rows and 7 columns. We save this new table as **Medalists since 1960**.

> You can similarly select part of a larger data table stored on a SAS Metadata server. Though not a database management system, the logic illustrated here still applies. To access the functionality, establish a connection to your SAS server, and then selecct **File ▶ SAS ▶ Browse Data**. The interface is intuitive and flexible.

If our goal is to examine medals won by different countries, we should note that this table is not quite useful in its current form. Each row represents one athlete. For individual sports, this serves our purpose. But, for team sports, a medal is awarded to each team member. If our goal is to count medals by country, this will inflate the count substantially. Instead, we'd need to count medals per event. We can accomplish this by constructing a slightly more complex query, which we can accomplish with Query Builder.

Querying a Database from JMP

Query Builder

Before venturing into examples of using **Query Builder**, a brief introduction to the tool is in order. Long-time JMP users might be unfamiliar with it because it is a relatively new feature of JMP. Using the tool, a user connects to the database, specifies tables needed for the query, and then chooses the specific data to satisfy the analytic goal. The interface provides an intuitive way to construct queries within a JMP session, rather than writing a query within the database environment and then exporting the result to a format that is readable by JMP.

In addition to generating code to extract data from a database, versions of **Query Builder** appear in two other JMP contexts. There is a **Query Builder** in the **File ▶ SAS** platform for selectively obtaining data from a SAS metadata server. That version is used in a short example at the end of this chapter. There is also a version on the **Tables** menu for building queries on JMP data tables, as opposed to database tables. That version is illustrated in Chapter 8.

There are three ways to launch **Query Builder** to query a relational database system:

- On the **File ▶ Database** menu, as shown earlier in Figure 5.1.
- In Windows, there is a **Query Builder** icon on the toolbar (Figure 5.14):

Figure 5.14: Query Builder Icon on Toolbar

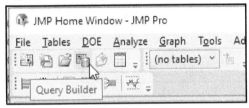

- At the beginning of a session, there is a launch button on the first page of **JMP Starter**, as shown here in Figure 5.15.

Figure 5.15: Query Builder Launch from JMP Starter

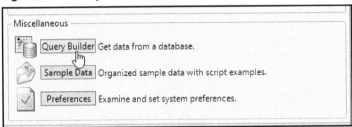

However you initiate **Query Builder**, the initial screen asks you to connect to a database. If you use both Query Builder and Open Table, any connections that you establish are available on both platforms. We'll run through two examples in this section, the first using the Olympics database, and the second using the Adventure Works database. With the Olympic medals query, we'll briefly illustrate selection of columns from a single database table with aggregation and relatively little commentary. The fuller explanation accompanies the bicycle production example. In both instances, we begin by selecting one or more tables for the query.

Revisiting the Olympic Medals Query

For this query, we again want to select the Medalist table from the Olympics schema, and then click **OK**. (See Figure 5.16.) This opens another dialog that enables us to join the primary table with one or more others.

Figure 5.16: First Step in Query Builder

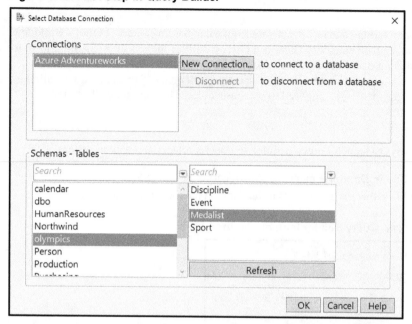

1. The **Medalist** table contains everything we'll need, so click **Build Query** to move on to choosing columns, specifying calculations and filters, as well as row ordering.

2. At the top of the dialog, name this query Summer Medals since 1960.

3. We now select the columns that we want (Figure 5.17): Event, NOC, Gender, Medal, and Season. These are the same as before. Highlight them in the **Available Columns** panel and click **Add**.

Figure 5.17: Specifying Query Conditions

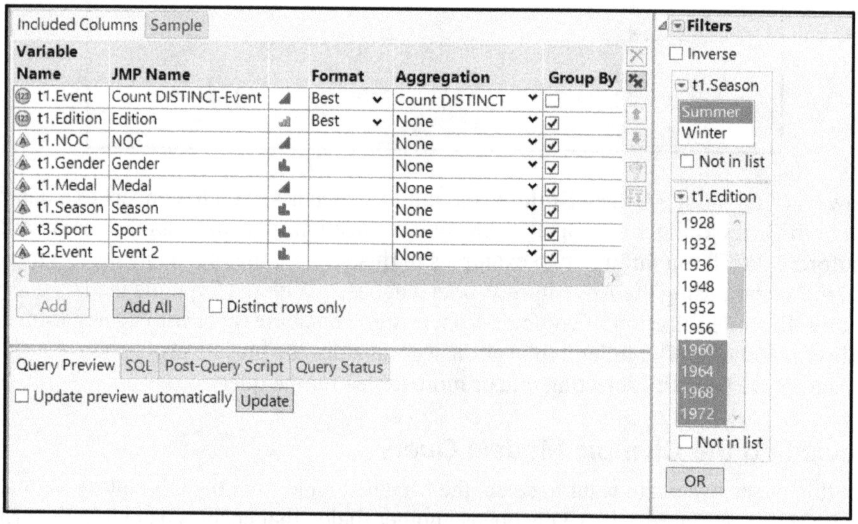

4. Make the following changes to the Included Columns in the upper middle of the dialog.

   a. In the **Aggregation** column for first row (**t1.Event**), click **None**, and change it to **Count DISTINCT**. For team events, this will count the team medals as 1 per medal rank (gold, silver, bronze) per event. Then change the **JMP Name** to **NMedals** for clarity.

   b. Select **t1.Edition**. In the untitled column between **JMP Name** and **Format**, click and change from **Continuous** to **Ordinal**, to switch the **Edition** data type.

   c. Highlight **t1.Season** and click the Filter icon that looks like a funnel. Alternatively, right-click and choose **Filter By**. This places an item in the Filters panel; choose Summer.

   d. Highlight **t1.Edition**, and add it to the Filters panel as well. Select all years from 1960 onward.

   e. To preview the query results, click **Update** near the bottom of the dialog. Then click **Run Query**.

This creates a considerably smaller table of 2,292 rows and 6 columns. The principal difference between this and the prior imported table is that now the number of medals won in different events are combined. So, for example, rows 3 through 8 of the result table show that Australian (AUS) men won bronze medals in 5 events, gold medals in 7, and silver in 5 during the 1960 games. The respective counts for women were 1 bronze, 1 gold, and 3 silver. We will return to this table in chapter 8. We'll now move on to a more comprehensive example that requires joining tables.

An Illustrative Scenario: Bicycle Parts

An analyst imports data from a database because she expects the selected data to serve her investigative purposes. In other words, the process of query-building happens within a problem-solving context. Hence, for this example, we start with a business problem.

The Microsoft Technet website (2016) provides the following background for the products made and sold by Adventure Works:

As a bicycle manufacturing company, Adventure Works Cycles has the following four product lines:

- Bicycles that are manufactured at the Adventure Works Cycles company.
- Bicycle components that are replacement parts, such as wheels, pedals, or brake assemblies.
- Bicycle apparel that is purchased from vendors for resale to Adventure Works Cycles customers.
- Bicycle accessories that are purchased from vendors for resale to Adventure Works Cycles customers.

Adventure Works Cycles (AWC) bikes are assembled using more than 500 different components such as pedals, chains, wheels, and so on. The firm manufactures some components itself, and purchases some from other vendors. In the following example, we'll focus on components, and leave the bicycles, apparel, and accessories for another time.

The database groups products into categories, each of which consists of multiple subcategories. The **Product** table includes a **ProductSubcategory** column. The **ProductSubcategory** code value for a product can then be used to look up the corresponding name of the subcategory and the parent category ID within the **ProductSubcategory** table. We can then use the category ID to find the category name in the **ProductCategory** table.

Suppose that an analyst in the firm wants to investigate the comparative profitability of the components that AWC makes and those that they buy. In addition to the direct dollar figures, she intends to account for the implicit costs associated with the time required to manufacture and the vendor lead time for those items bought.

The logic of the query will be as follows: We want to import data about items in the components category. As we see in Table 5.1 below, the **Product** table has a subcategory column, but no category column. However, subcategories are associated with categories so that we can find the rows we want by using both the **Category** and **Subcategory** tables.

The relevant columns reside in three tables within the **Production** schema and one table in the **Purchasing** schema. The two schemas combined have more than thirty tables and several hundred columns, so it will be efficient to extract only the data needed for this study. Table 5.1 identifies the tables and the relevant columns, noting which columns are Primary Keys (PK), Foreign Keys, and Unique Key (indexed) Columns.

Table 5.1: Components Data to Import from the Adventure Works Database

Table	Column Name	Key
Production.Product	ProductID	PK
	Name (of product)	UK
	MakeFlag (0-1 buy or make)	
	StandardCost	
	ListPrice	
	DaysToManufacture	
	ProductSubcategoryID	FK
Production.ProductSubcategory	ProductSubcategoryID	PK
	ProductCategoryID	FK
	Name (of subcategory)	UK
Production.ProductCategory	ProductCategoryID	PK
	Name (of category)	UK
Purchasing.ProductVendor	ProductID	PK, FK
	BusinessEntityID (vendor code)	
	AverageLeadTime	
	StandardPrice	
	LastReceiptCost	

We should note also that there are three columns titled Name but they have different contents in this database. Bear this anomaly in mind as we move forward. With a clear sense of the columns of interest, we can now turn to **Query Builder** to construct the query.

Designing a Query with Query Builder

Query Builder follows a two-stage process. In the first, we choose the tables that we'll join for access to the columns of interest. In the second stage, we choose the columns (variables) and specify conditions. In this example, we just want the columns listed in Table 5.1 where **Name (of category)** = Components.

The first phase is accomplished via the **Select Tables for Query** dialog (Figure 5.18). Table selection requires a **Primary** table and can include any number of **Secondary** tables. Eric Hill, creator of **Query Builder**, explains the distinction this way (2015):

The **Primary** table in a query is normally the table from which all rows should be retrieved, regardless of whether they have a matching value in some other table. This is your **fact table**, in data warehousing terms. There can only be one primary table in a query.

Secondary tables are normally lookup tables. By default, rows from secondary tables are only included in the query when the value of a column in the primary table matches the value of a specified column in the secondary table. These are your **dimension tables**, in data warehousing terms. You can add as many second tables to the query as you want.

In this example, the Production.Product table is the primary table, and the other three will be the secondary tables.

1. In the upper left of the dialog in Figure 5.18, make sure that the Production schema is selected. Highlight Product in the **Available Tables** list, and click **Primary** under **Select Tables for Query**.

2. Look at the **Columns** panel in the lower right portion of the dialog, which displays the structure and content of the selected table. In particular, the Production.Product table has 504 rows and 25 columns. Each column is listed along with its data type. Key columns are identified with one of three icons corresponding to primary (gold), foreign (blue with a pale blue shadow), or unique keys. In the case of foreign keys, the **Reference** column identifies the corresponding table. These reference tables are candidate secondary tables.

3. In the same lower right portion, click the **Table Snapshot** tab to preview the data values in the table. This is a preview of the data available for import.

4. After scanning the data, click the **Column** tab again.

Figure 5.18: Selecting Tables for the Query

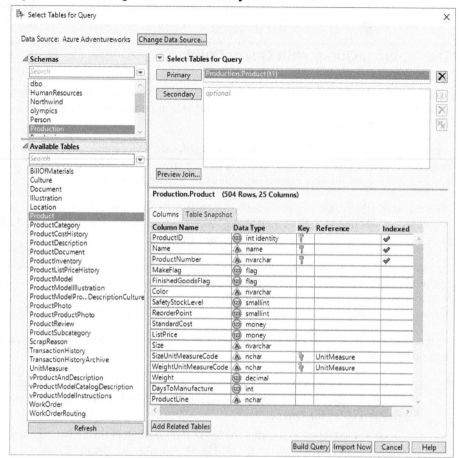

We want to add three secondary tables. **Query Builder** offers a shortcut via the **Add Related Tables** button at the bottom of the dialog. Clicking the button adds all referenced tables to the list of Secondary tables at the top of the dialog. In this illustration, however, we want some but not all of the tables included, so we'll select them individually.

5. The two other required tables in the **Production** schema are **Product.Category** and **Product.Subcategory**. Select both tables (holding down the **Ctrl** key to multi-select) and then click **Secondary**.

6. The third secondary table is in the **Purchasing** schema. To add it to the list, we first choose **Production** in the **Schemas** list, and then highlight **ProductVemdor** among **Available Tables** and click **Secondary**.

After making the selections, the **Select Tables** zone looks like Figure 5.19. At this stage, JMP is initially planning to merge the tables with outer joins among pairs of tables in our list, as evidenced by the small Venn diagram symbol next to each secondary table. Query Builder recommends potential joins based on the information available, which might or might not be what we want. Hence, it behooves us to inspect and possibly edit the joins.

Figure 5.19: List of Tables to Join

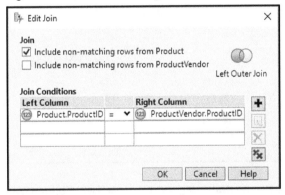

7. Highlight each secondary table in sequence, and click **Edit Join** 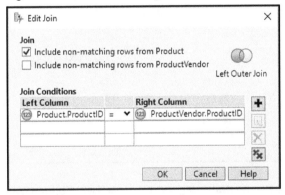 in the outer margin.

Figure 5.20 shows the proposed join between **Production.Product** and **Purchasing.Product Vendor**. This is the correct association: All **ProductID** codes from the **Product** table will be looked up in the **VendorProduct** table.

Figure 5.20: Edit Join Dialog

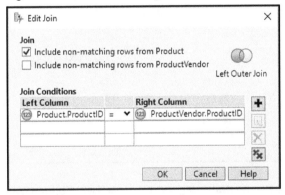

Similarly, JMP correctly anticipates the join between the **Product** table and **ProductSubcategory** tables. However, the association with the **ProductCategory** table (t2) is incorrect and needs modification. The **Edit Join** dialog (Figure 5.21) reveals the issue. The JMP best guess is to match the **Name** columns in the two tables, but our intention is to join **ProductCategory** and **ProductSubcategory** tables on the common **ProductCategoryID** columns.

Figure 5.21: A Potential Join in Need of Adjustment

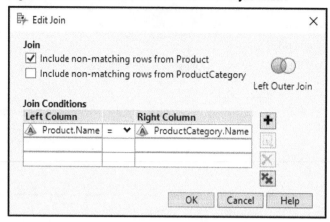

8. To correct the **Join Conditions**, highlight the only row and click the **Edit** icon [icon] (below the **Add** icon) to open the Edit Condition dialog (Figure 5.22).

Figure 5.22: Editing a Join Condition

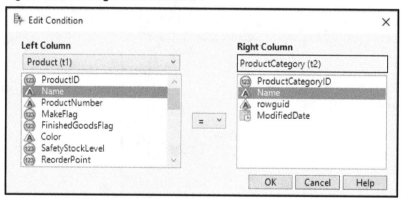

9. On the left side, change **ProductSubcategory (t3)** as the **Left Column** table and **ProductCategoryID** as the column to join.

10. On the right side, change **ProductCategoryID** as the column to join and click **OK**.

11. This returns us to the **Select Tables for Query** dialog, where we find a **Preview Join** button. Clicking this button presents a preview of the resulting query table at this stage in the process. It is a good idea to see the proposed join to confirm that we have what we intended.

    a. If further corrections are needed, go back and add or remove secondary tables or adjust the join conditions.

    b. Once everything is in order, click **Build Query** to open the **Query Builder** window as shown in Figure 5.23.

Figure 5.23: Main Query Builder Dialog

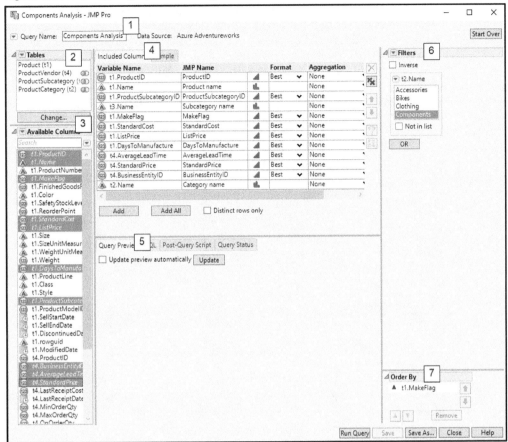

Given the complexity of this dialog, let's first review the seven boxes it contains to see how we create the query.

1. The **Query Name** box enables us to name the query for saving and future reference. By default, JMP uses the primary table name, but it is wise to give the query a more meaningful name. I have chosen Components Analysis here. We also find the **Data Source** here as well.

2. The **Tables** box summarizes the work that we did in the prior dialogs. Click **Change** if you need to return to those dialogs.

3. The **Available Columns** box lists all columns from the primary and secondary tables as well as reference to the source tables (t1, t2, and so on). Note that joined columns appear repeatedly in this list—once for each source table. In Figure 5.23, I have selected several columns and plan to click **Add** in the **Included Columns** box (# 4). You can also click and drag column names there or to the **Filter** and **Order By** boxes (#6 & #7). Right-clicking on a column name opens options to accomplish the same ends. After you have included a column in the query, its name becomes italicized in the **Available Columns** list.

4. The fourth box has two panels. The **Included Columns** panel displays a table of selected columns along with editable attributes. In the grid, we can rename columns and modify

modeling types and formats before importing into JMP. (Of course, these can be changed at any later time.) For example, three of the four source tables have a column titled Name. I have revised those as Product name, Subcategory name, and Category name. We can also continue to add, remove, or reorder columns in the list, or designate any of them as **Filter** or **Order By** columns using the buttons to the right. Figure 5.24 shows the panel after we selected columns and edited some attributes.

Figure 5.24: Included Columns Panel with Customized Column Attributes

Variable Name	JMP Name	Format	Aggregation	Group By
t1.ProductID	ProductID	Best	None	☐
t1.Name	Product name		None	☐
t1.ProductSubcategoryID	ProductSubcategoryID	Best	None	☐
t3.Name	Subcategory name		None	☐
t1.MakeFlag	MakeFlag	Best	None	☐
t1.StandardCost	StandardCost	Best	None	☐
t1.ListPrice	ListPrice	Best	None	☐
t1.DaysToManufacture	DaysToManufacture	Best	None	☐
t4.AverageLeadTime	AverageLeadTime	Best	None	☐
t4.StandardPrice	StandardPrice	Best	None	☐
t4.BusinessEntityID	BusinessEntityID	Best	None	☐
t2.Name	Category name		None	☐

The second panel is accessed on the **Sample** tab. In this panel, you can generate a random sample if the server's database supports random sampling, or just the first *n* rows of data. Also, the panel allows for setting a random number seed for repeatable results.

5. The lower central portion of the dialog features five tabs:

  ○ **Query Preview** displays the data that the query will return when run. The **Update** button refreshes the preview as we alter the specifications.

  ○ **SQL** shows the current state of the generated SQL code.

  ○ **Post-Query Script** provides the option to include a JSL script to run after the query has completed.

  ○ **Query Status** lists the current status of other currently running queries, along with an option to stop them.

6. We specify column-based criteria to filter rows in the query in the **Filter** panel. In terms of SQL, these will become the conditions in the WHERE clause. In this example, we want to filter by Category ID, and download only rows for category #2. (See left panel of Figure 5.25.) Note that the red triangles in this panel offer several options, including the option to prompt the user for a filter value when the query runs.

7. Finally, the **Order By** panel is where the elements of the SQL Order By clause are specified. In this query, we'll order the rows first by buy-or-make status, and then by subcategory number, as shown in the right panel of Figure 5.25.

Figure 5.25: Filter and Order by Conditions

After all aspects of the query have been specified, it's time to **Save** the query specifications (for reproducibility) and to **Run Query**.

The query result appears like any JMP table. It has four table variables similar to earlier import examples that we have seen. The first table variable is called SQL, and it is the SQL code generated by our work in Query Builder (Figure 5.26).

Figure 5.26: The Final SQL Code

```
Table Variable for Components Analysis - JMP Pro                    —   □   ✕

Name SQL                                                              OK

Value   SELECT t1.ProductID, t1.Name AS [Product name],              Cancel
        t1.ProductSubcategoryID, t3.Name AS [Subcategory name],
            t1.MakeFlag, t1.StandardCost, t1.ListPrice,
        t1.DaysToManufacture,
            t4.AverageLeadTime, t4.StandardPrice, t4.BusinessEntityID,
        t2.Name AS [Category name]
        FROM Production.Product   t1
            LEFT OUTER JOIN Purchasing.ProductVendor t4
                ON ( t1.ProductID = t4.ProductID )
            LEFT OUTER JOIN Production.ProductSubcategory t3
                ON ( t3.ProductSubcategoryID =
        t1.ProductSubcategoryID )
            LEFT OUTER JOIN Production.ProductCategory t2
                ON ( t3.ProductCategoryID = t2.ProductCategoryID )
        WHERE ( ( ( t2.Name IN ( 'Components' ) ) ) )
        ORDER BY t1.MakeFlag ASC;
```

The other table variables are JSL scripts, which are analogous to the ones that we have seen earlier. After we save the data table within JMP, those scripts are preserved for future use or modification.

Query Builder has considerably more functionality than demonstrated in this example, but this should help orient you to the platform and to begin designing queries. Experienced database users might wish to write or edit their own code, and **Query Builder** can accommodate that preference. Also, it is possible to aggregate data and perform some preliminary analysis. All of these capabilities are documented within **Help** and the Hill white paper (2015).

Query Builder for SAS Server Data

As in Chapter 4, we now return to an example from the Advanced Business Analytics course (Truxillo 2012) designed by SAS. In this illustration, we have transactional data from a financial institution. The institution wants to extract, transform, and load data to develop a model to "identify a subset most likely to have interest in an insurance investment product" (pp. 2-26).

The fundamental data model is a star schema, with a central fact table and several dimensional tables. For this project, the strategy will be to assemble a set of columns about clients with and without insurance products. Because there are far more customers without insurance than with, we'll deliberately undersample those without, so that the final query table contains comparable numbers from each of the two target categories.

Specifically, there are five linked tables involved here: clients, client insurance accounts, insurance accounts, credit bureau, and non-insured qualified clients. For this example, we'll join the four complete tables as part of the larger project, beginning with the list of insurance accounts. Then we'll select a random subset of the non-insured clients.

1. Select File ▶ **SAS** ▶ **Query Builder** and connect to your SAS server.
2. After you are connected, select the appropriate schema and choose a Primary table from the list of available tables. Here, we choose **ABA1** as the schema and **INS_ACCOUNT** as the primary table (Figure 5.27). Notice the similarities between this dialog and the dialog in Figure 5.17. This table contains just columns related to all insurance accounts.

Figure 5.27: Selecting Tables for Query

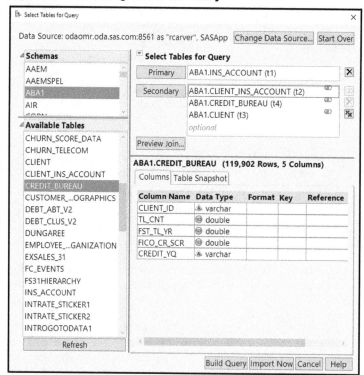

3. Now select the table **CLIENT_INS_ACCOUNT** as a **Secondary** table to link customers to these insurance accounts. Note that the tiny Venn diagram icon just to the right of **ABA1.CLIENT_INS_ACCOUNT(t2)** indicates a proposed left outer join on the common **INS_ACT_ID** column, which is correct. Click **Preview Join** to confirm.

4. We continue by selecting **CLIENT** as a secondary table, which contains some basic information about clients. This will be joined on the **CLIENT_ID** column that it shares with the **CLIENT_INS_ACCOUNT** table.

5. Lastly, we add the **CREDIT_BUREAU** table to the secondary list. This table also has a **CLIENT_ID** column. Then click **Build Query**.

   At this point, the query-building process has access to all columns and rows in the five tables combined. Depending on knowledge of the business, we might filter insurance accounts and/or clients to restrict the data extraction by dates or some other factor. We would do so by setting filters just as in the Adventure Works example earlier. For now, we won't filter these tables at all.

6. In the main **Query Builder** window (Figure 5.28), we first name the query at the top of the screen. Here I've named it **INS_CLASS** for insurance classification data. Note that I have resized the dialog and its panels for clarity. Before the screen capture for Figure 5.28, we clicked **Add All** to select all of the columns from those tables. This will provisionally include repeated columns, such as **CLIENT_ID**, but we can remove those columns individually before running the query. For this illustration, we should **Save** the query specification and click **Run Query**.

Figure 5.28: Selecting Columns and Rows for the Query

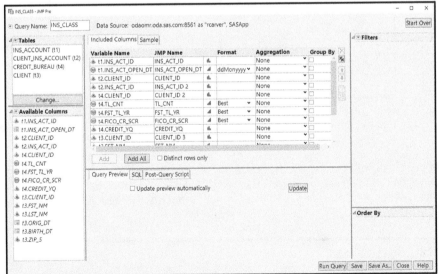

The resulting JMP data table consists of 15 columns and 2,028 rows. As before, embedded within the data table is the SQL code and JSL scripts that enable us to rerun or modify the query at a later time.

7. Next, we move on to select a subset of 2,028 individuals from the **NON_INS_CLIENTS_QUALIFIED** table so that we'll have a similar number of clients with and without policies. We can do this in several ways, but we'll continue with **Query Builder**. First, click **Start Over** in the **Query Builder** dialog.

8. This reopens the dialog from Figure 5.27. Just choose the one table **NON_INS_CLIENTS_QUALIFIED** and click **OK.**

9. In the next dialog, rename the query to a meaningful name, such as **NON_INS_2028**, and add all columns.

10. Now click the **Sample** tab in the central panel of the dialog.

11. In the sample panel, we have checked the box next to **Sample this result set**, specified a sample of *n* = 2028 rows, and set a random number seed of 16149 (Figure 5.29).

Figure 5.29: Drawing a Random Sample of Rows from a Query Result Set

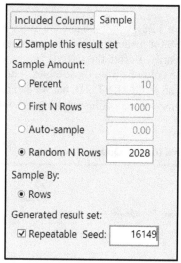

12. Finally, run the query to generate a table of qualified clients who have not yet bought an insurance product and save the resulting JMP data table.

At this point, we've extracted all the data we want from the SAS server. We can close the connection and prepare to combine the two JMP data tables. We'll see how to do that in Chapter 8 after treating some other common data extraction and transformation issues.

Conclusion

In this chapter, we have seen several methods for extracting database data and converting it to *.jmp formatted files. This is the second of three chapters dealing with data importation. The final area deals with data accessed directly from a website, and that is the subject of the next chapter.

References

Hill, Eric. 2015. SAS Institute white paper. "Query Builder: The New JMP 12 Tool for Getting Your SQL Data into JMP." Available at https://www.jmp.com/content/dam/jmp/documents/en/white-papers/query-builder-jmp-12-107669_0515.pdf.

Microsoft TechNet. 2016. "Adventure Works Cycles Business Scenarios: Product Scenario." Available at https://technet.microsoft.com/en-us/library/ms124670(v=sql.100).aspx.

PostgreSQL Global Development Group. 2016. Dellstore2 Sample Database. Available at https://wiki.postgresql.org/wiki/Sample_Databases.

SAS Institute Inc. 2016. *Using JMP 13*. Online book provided with JMP Help Menu.

Truxillo, Catherine. 2012. *Advanced Business Analytics Course Notes, Volume 1*. Cary NC: SAS Institute Inc.

Chapter 6: Importing Data from Websites

Introduction

The previous two chapters have covered methods for importing data from internal sources—that is, data that resides in your own system or data warehouse. For many projects, the data that we already have are insufficient for the analytic purpose. Professor Chris Nachtsheim (2016) of the University of Minnesota attributes the "Fayyad axiom" to Usama Fayyad, Chief Data Officer of Barclays and co-founder of the KDD conferences: "Your organization's data warehouse will not have the information you need to build your predictive model." As we've noted before, it is increasingly common for analytic projects to weave together data from multiple sources, and commonly you might seek publicly available data from the Internet.

Chapter 4 used the example of data downloaded in Excel and CSV formats from websites. JMP also has the ability to recognize tabular data embedded in a web page and to read that table directly into a JMP data table. This brief chapter explains how to import data directly from a URL or an FTP site. The command that does this will also import from a remote computer, but we've touched on the latter operation in Chapter 5.

In Chapter 5, we imported data about all Olympic medalists since 1900. In later chapters, we will want to investigate the comparative success of countries in winning Olympic medals, but will need some additional data for a meaningful study. Why? The medalist data include all countries that have sent winning athletes to the games, but omit unsuccessful participating nations. As such, we'll want to find a list of all nations that participate in the modern games. Such a list is easy to locate, and we'll want to use the JMP **Internet Open** command to transfer the data.

In this chapter, our main focus will be on transferring a table from a web page directly into a JMP data table. Once a user understands how to perform the operation with a URL, it's a short step to doing the same thing with paths starting with FTP or with the address of a remote computer on your network.

Variety of Web Formats

At the outset, it is important to recognize that there are several standards in use to embed and present tables online. JMP will import data from many of these, but not all. There are JMP Add-ins available for some specific sites such as Yahoo! Finance (Liu 2014), and other add-ins might be created after publication of this book. Though it is beyond the scope of this book, more advanced readers might want to write JSL scripts to manage the extraction task for non-supported formats, and the JMP Community site (community.jmp.com) is an excellent starting point.

Internet Open

To supplement the table of Olympic medal winners, we want to create a table containing all nations that participate in the Olympic games. A quick search for a definitive list of IOC countries led me to Wikipedia, specifically to https://en.wikipedia.org/wiki/List_of_IOC_country_codes, with a detail of the result shown in Figure 6.1.

Figure 6.1: IOC Country Codes

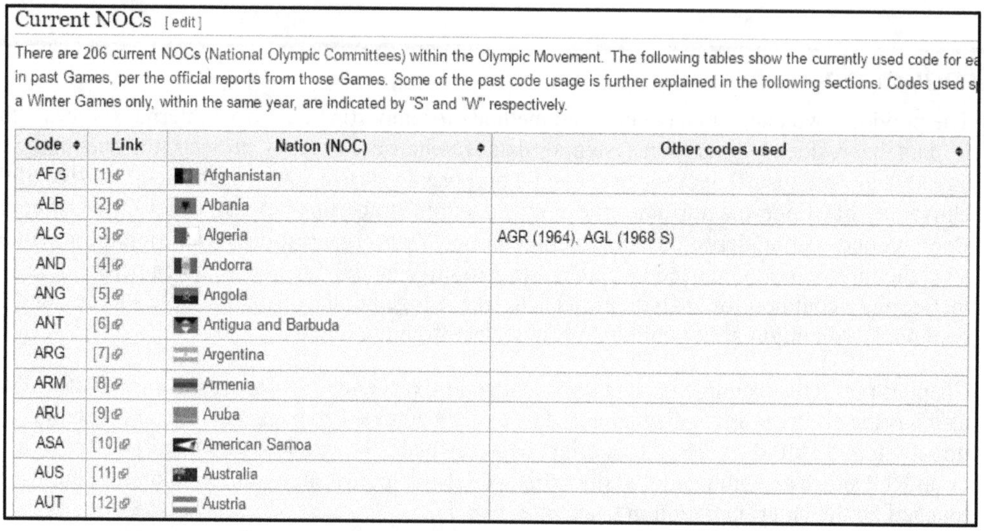

Current NOCs [edit]

There are 206 current NOCs (National Olympic Committees) within the Olympic Movement. The following tables show the currently used code for ea in past Games, per the official reports from those Games. Some of the past code usage is further explained in the following sections. Codes used s a Winter Games only, within the same year, are indicated by "S" and "W" respectively.

Code ◦	Link	Nation (NOC) ◦	Other codes used ◦
AFG	[1]⧉	Afghanistan	
ALB	[2]⧉	Albania	
ALG	[3]⧉	Algeria	AGR (1964), AGL (1968 S)
AND	[4]⧉	Andorra	
ANG	[5]⧉	Angola	
ANT	[6]⧉	Antigua and Barbuda	
ARG	[7]⧉	Argentina	
ARM	[8]⧉	Armenia	
ARU	[9]⧉	Aruba	
ASA	[10]⧉	American Samoa	
AUS	[11]⧉	Australia	
AUT	[12]⧉	Austria	

When you scroll through the page, you see that both country names and NOC codes have changed over the years. On this page, there are actually three tables of NOC codes: one with the 206 nations that currently compete, one with 14 country names for historic reference to past winners, and a set of obsolete codes for countries whose names have changed. For example, the nation known as BIR Burma competed from 1948-1988 and is now known as MYA Myanmar. For the sake of this example, we'll focus on the first table of current codes to illustrate the approach, but will import the three relevant tables to acknowledge that it will be important to discover and reconcile historic and obsolete codes should any occur in the table of medal winners.

To quickly open the data as a JMP table, select **File ▶ Internet Open** and first specify the URL (Figure 6.2). Copy and paste the URL into the dialog as shown, and specify that you want to open is as **Data**.

Figure 6.2: Internet Open

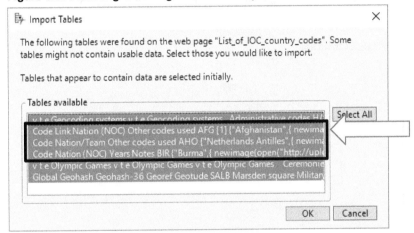

Next comes a dialog (Figure 6.3) showing the JMP "best guess" about the presence of tabular data on the particular web page. In addition to the three tables of codes, there are some parts of the page that might not strike the viewer as tables. The list of **Tables available** will typically display the contents of the first couple of row. In this instance, we recognize that we want the data residing in the second, third, and fourth tables, so just change the highlighting and click **OK**.

Figure 6.3: Selecting Web Page Tables to Import

This immediately opens a new data table, a close-up of which is shown in Figure 6.4. It consists of four Nominal columns and 206 rows, corresponding to the 206 IOC nations. Also notice that the second column has embedded code referring to the name of the country and the URL for the national flag. Notice further that there are two scripts among the table variables (upper left; highlighted in Figure 6.4) titled **Source** and **Load** pictures for "Nation (NOC)". Clicking the green arrow runs the script, and right-clicking on the name of the script offers the option to **Edit** (and thus view) the script. The "**Source**" script is a record of the **Internet Open** command that produced the table in the first place. Running it again re-opens the table, which might be helpful in the future should the country list change in any way. Also, once the data table is saved, the script is preserved.

Figure 6.4: The First Imported Table

		Code	Link	Nation (NOC)
	1	AFG	[1]	{"Afghanistan",{newimage(open("http://upload.wikimedia.org...
	2	ALB	[2]	{"Albania",{newimage(open("http://upload.wikimedia.org/wiki...
	3	ALG	[3]	{"Algeria",{newimage(open("http://upload.wikimedia.org/wiki...
	4	AND	[4]	{"Andorra",{newimage(open("http://upload.wikimedia.org/wik...
	5	ANG	[5]	{"Angola",{newimage(open("http://upload.wikimedia.org/wiki...
	6	ANT	[6]	{"Antigua and Barbuda",{newimage(open("http://upload.wiki...
	7	ARG	[7]	{"Argentina",{newimage(open("http://upload.wikimedia.org/w...
	8	ARM	[8]	{"Armenia",{newimage(open("http://upload.wikimedia.org/wi...
	9	ARU	[9]	{"Aruba",{newimage(open("http://upload.wikimedia.org/wikip...
	10	ASA	[10]	{"American Samoa",{newimage(open("http://upload.wikimedi...
	11	AUS	[11]	{"Australia",{newimage(open("http://upload.wikimedia.org/wi...
	12	AUT	[12]	{"Austria",{newimage(open("http://upload.wikimedia.org/wiki...

List of IOC country codes 2 — Source — Load pictures for "Nation (NOC)" — Columns (4/1): Code, Link, Nation (NOC), Other codes used

Running the second script adds a new column to the far right of the data table (Figure 6.5). For later visualizations, it might be handy to have those flag images as an Expression data type in our table.

Figure 6.5: National Flags Unfurled

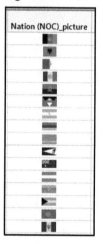

Nation (NOC)_picture

Since the NOC code and flag are really the two useful columns, let's delete the other columns and rename the **Nation (NOC)_picture** as Flag. At this point, you should save the data table with a suitable name. I have placed a copy of the file named **List_of_IOC_ country_codes 2** in the list of files that accompany the book.

That's really all there is to the import process. You might repeat the process for the remaining two tables, as needed. In Chapter 8, we'll see how we might combine the three files and/or join them with the medalist data.

Common Issues to Anticipate

As noted earlier, you might periodically find a website for which **Internet Open** will not work. Some websites dynamically load a table after the web page itself loads. On such sites, the data are not embedded within the code that creates the page, but rather the page contains code to retrieve the data each time the page reloads. The Windows versions of JMP offer a workaround; this is not available in Macintosh versions as of this writing.

To illustrate, we'll visit a website maintained by Major League Baseball in the US. This particular page provides a leaderboard for the longest home runs recorded in a given season. The URL is http://m.mlb.com/statcast/leaderboard#hr-distance. See Figure 6.6.

Figure 6.6: MLB Home Run Distance Leaderboard for the 2016 Season

Now within JMP, do the following:

- Select **File ▶ Internet Open** Enter the URL and leave the default setting **Open As: Data**.
- Click **OK**, which generates the alert shown in Figure 6.7.

Figure 6.7: Internet Open Alert

> JMP Alert   ✕
>
> ❓ No tables found.
>
> This web page is not formatted in a way that allows its data to be imported.
>
> OK

- Click **OK** in the alert, and then again select **File ▶ Internet Open**. Enter the URL and this time change the default to **Open As: Web Page**.
- This opens a new web browser within the JMP session. Within the new browser window, select **File ▶ Import Table as Data Table**. As we saw in Figure 6.3, JMP makes its best guess about the available data tables.

- Because it is not obvious which table is the one we want, just click **OK**.

In this instance, JMP opens four tables. It happens that, for this page, the first table contains all of the data displayed on the website. The imported table is shown in part as Figure 6.8.

Figure 6.8: The Imported MLB Leaderboard Data

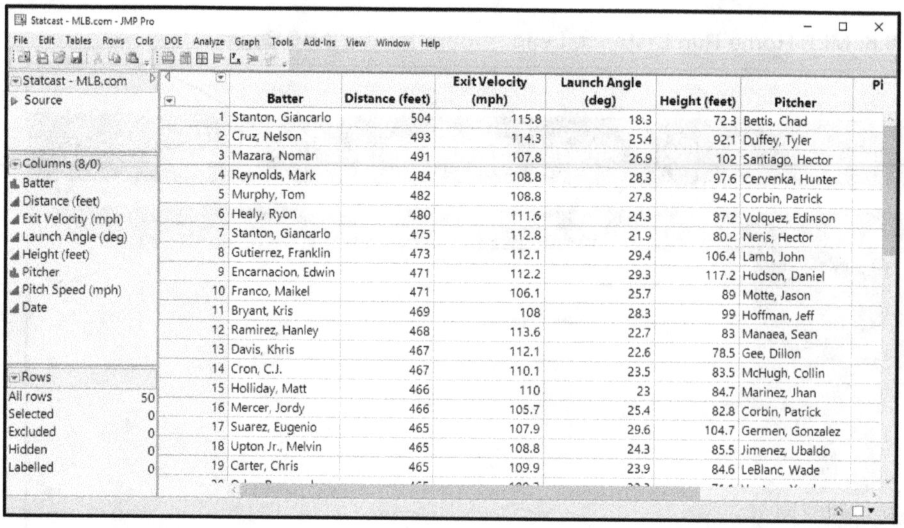

	Batter	Distance (feet)	Exit Velocity (mph)	Launch Angle (deg)	Height (feet)	Pitcher
1	Stanton, Giancarlo	504	115.8	18.3	72.3	Bettis, Chad
2	Cruz, Nelson	493	114.3	25.4	92.1	Duffey, Tyler
3	Mazara, Nomar	491	107.8	26.9	102	Santiago, Hector
4	Reynolds, Mark	484	108.8	28.3	97.6	Cervenka, Hunter
5	Murphy, Tom	482	108.8	27.8	94.2	Corbin, Patrick
6	Healy, Ryon	480	111.6	24.3	87.2	Volquez, Edinson
7	Stanton, Giancarlo	475	112.8	21.9	80.2	Neris, Hector
8	Gutierrez, Franklin	473	112.1	29.4	106.4	Lamb, John
9	Encarnacion, Edwin	471	112.2	29.3	117.2	Hudson, Daniel
10	Franco, Maikel	471	106.1	25.7	89	Motte, Jason
11	Bryant, Kris	469	108	28.3	99	Hoffman, Jeff
12	Ramirez, Hanley	468	113.6	22.7	83	Manaea, Sean
13	Davis, Khris	467	112.1	22.6	78.5	Gee, Dillon
14	Cron, C.J.	467	110.1	23.5	83.5	McHugh, Collin
15	Holliday, Matt	466	110	23	84.7	Marinez, Jhan
16	Mercer, Jordy	466	105.7	25.4	82.8	Corbin, Patrick
17	Suarez, Eugenio	465	107.9	29.6	104.7	Germen, Gonzalez
18	Upton Jr., Melvin	465	108.8	24.3	85.5	Jimenez, Ubaldo
19	Carter, Chris	465	109.9	23.9	84.6	LeBlanc, Wade

If this workaround is not successful, or if you are using a Macintosh, it would be wise to consult the JMP user community site for workarounds. Sometimes you can highlight the desired data, copy, and paste it into a new worksheet. That might subsequently require further edits and formatting, but might do the job.

As with other imports described in Chapters 4 and 5, always verify data and modeling types. Numeric, character, and data/time data are generally quite straightforward, but images and hyperlinks warrant closer attention. The **Internet Open** approach manages the latter quite well, but copy and paste will typically fail to transfer the links. For unusual situations, consult the JMP documentation, especially *Using JMP*, Chapter 3: "Import Your Data."

Conclusion

This short chapter has illustrated the use of a powerful and simple tool for extracting data from a website. This tool can handle the usual complement of data types and truly can speed this particular type of dirty work on websites that do not themselves offer a data download option.

Users whose needs are more sophisticated than those demonstrated here or who have scripting experience should also investigate the expanded capabilities of writing your own JSL script. Hecht (2014) is an excellent starting point.

Chapters 4, 5, and 6 have all shared the objective of extracting data from one location or format into a single JMP data table. The next two chapters cover methods for reshaping a table and for consolidating data from several tables.

References

Hecht, Michael. 2014. "Web scraping with JMP for fun and profit". *SESUG 2014: The Proceedings of the SouthEast SAS Users Group.* Cary, NC: SAS Institute Inc.. Downloaded from http://analytics.ncsu.edu/sesug/2014/RIV-09.pdf.

Liu, Peng. 2014. "Yahoo Finance Fetcher." JMP Add-Ins. Cary, NC: SAS Institute Inc. Downloaded from https://community.jmp.com/t5/JMP-Add-Ins/Yahoo-Finance-Fetcher/ta-p/21509.

Major League Baseball. 2016. "Statcast Leaderboard: HR Distance." Downloaded 25 November, 2016, from http://m.mlb.com/statcast/leaderboard#hr-distance.

Nachtsheim, Christopher. 2016. "DOE: Is the Future Optimal?" *JMP Discovery Summit 2016, Plenary Session.* Cary, NC. Downloaded from https://community.jmp.com/t5/Discovery-Summit-2016/Plenary-Christopher-Nachtsheim/ta-p/23972.

SAS Institute Inc. 2016. *Using JMP 13*, Chapter 3. Cary NC: SAS Institute Inc.

Wikipedia. 2016. "List of IOC country codes." Downloaded on October 15, 2016, from https://en.wikipedia.org/wiki/List_of_IOC_country_codes.

Chapter 7: Reshaping a Data Table

Introduction

For most analyses, we organize data into a two-dimensional tabular format in which each row is an individual observation or case, and each column contains a measurement or attribute pertaining to that individual. There are times, though, when this is not the only option or even the preferred option. This is particularly the case in longitudinal or repeated-measures studies where we will have multiple values for the same measurement and the same subject. For example, in an investigation using repeated measurements of the same observational units, there are choices to be made about how to arrange the data. We can refer to the choices in terms of the *shape* of the data table.

As noted earlier in Chapter 2, we ordinarily strive for "tidy data" (Wickham 2014) that conform to the third normal form (Codd 1990) of relational databases:

1. Each variable forms a column.

2. Each observation forms a row.

3. Each type of observational unit forms a table.

What do we do when the third normal form does not meet the needs of a particular analysis?

What Shape Is a Data Table?

Consider a simple pilot study for a controlled experiment. The experiment investigates the effect of a training activity intended to increase a subject's score in performing a task. Suppose that the experiment is designed so that each subject in the study completes a pre- and post-test, and the scores are recorded. There are eight pilot subjects identified as individuals A through H, four of whom are randomly assigned to use the activity. The other four are the control group.

After data collection, we'll have two scores for each person, and we'll know which individuals were in each group. What are our choices for placing the data into a single matrix or data table?

Wide versus Long Format

If we follow the convention of allocating one row per experimental subject, our data table might look like this:

Table 7.1: Wide Array of Artificial Experimental Data

Subject	Group	Pre	Post
A	Exp	51	48
B	Ctrl	53	42
C	Ctrl	48	56
D	Exp	44	49
E	Exp	47	59
F	Ctrl	42	43
G	Ctrl	52	39
H	Exp	48	49

This format is often referred to as "wide": All of the attributes and measurements of a given individual are spread across several columns. In this example, we have a matrix of eight rows and four columns with minimal redundancy.

An alternative to the wide format is to re-arrange the same data in a "narrow" or "tall" arrangement. This introduces some redundancy in the **Subject**, **Group**, and **Test** columns in exchange for treating the test (pre- or post-) as a categorical variable and treating all of the scores as a single numeric variable.

Table 7.2: Long Arrangement of the Same Data

Subject	Group	Test	Score
A	Exp	Pre	51
A	Exp	Post	48
B	Ctrl	Pre	53
B	Ctrl	Post	42
C	Ctrl	Pre	48
C	Ctrl	Post	56
D	Exp	Pre	44
D	Exp	Post	49
E	Exp	Pre	47
E	Exp	Post	59

Subject	Group	Test	Score
F	Ctrl	Pre	42
F	Ctrl	Post	43
G	Ctrl	Pre	52
G	Ctrl	Post	39
H	Exp	Pre	48
H	Exp	Post	49

Ultimately, the decision about the preferred shape of a data table should be driven by the logic of your analytical goals. In designing an experiment or data collection from scratch, you should consider the analytical goals from the outset. Frequently, though, the layout of a data table was chosen at an earlier time by others, and the job of the analyst is to render the data into a format that suits the current purpose.

Reasons for Wide and Long Formats

Perhaps the most common reason for arranging data into a wide format is that we are working with longitudinal or repeated measures data. Some analysis platforms might require that each repeated measurement be treated as a variable. (See, for example, Grace-Martin 2015, or SAS Institute 2016a.) For some procedures, JMP can model repeated measures in either wide or long format, as discussed in SAS Institute 2016b. Because repeated measures of the same subject or observational unit are likely to be correlated with one another, they are better treated as multiple variables rather than a single variable.

When the analytical goal involves explicitly viewing time as a factor or predictor, as in time series modeling or visualization, then a long format might be more suitable. For example, a data table should be long to use the Time dimension on the JMP **Bubble Plot** platform.

The next two sections illustrate the JMP method for converting wide to narrow format and vice versa. The initial illustration uses this artificial data set only to present the method simply before moving on to more complex illustrations.

Stacking Wide Data

In this introductory example, the initial state of the data is wide with one row per subject and the two measurements occupying separate columns. In other words, the two measurements are being treated as two variables. Figure 7.1 shows the initial state of the data table. Because the design of this hypothetical study uses matched pairs, we might move directly to analysis with the data table.

For other analytical purposes, it could be helpful to reshape the table to the narrow format. We use the **Stack** command in the **Tables** menu.

Figure 7.1: Example Data Table in Wide Format

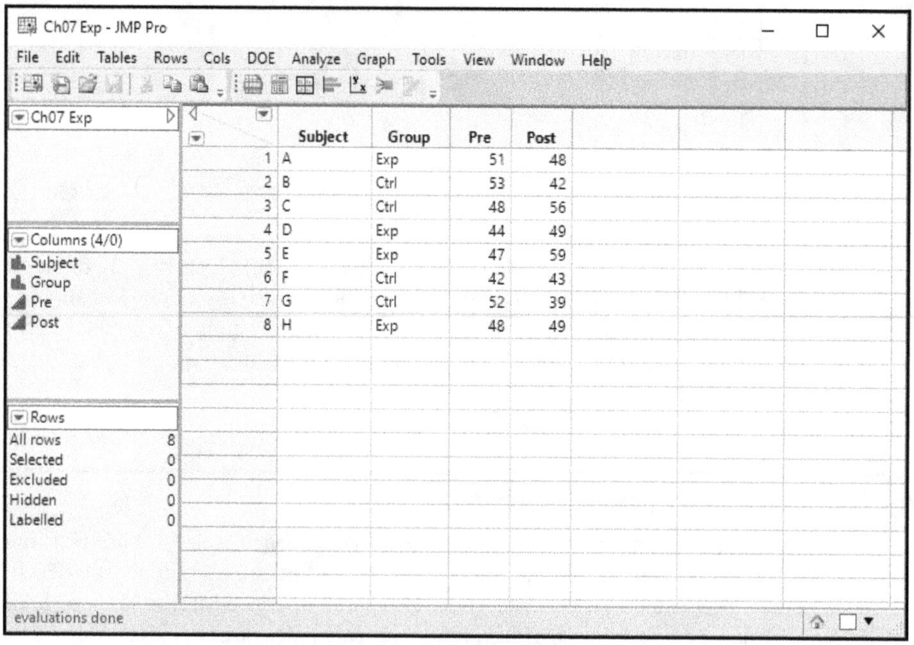

	Subject	Group	Pre	Post
1	A	Exp	51	48
2	B	Ctrl	53	42
3	C	Ctrl	48	56
4	D	Exp	44	49
5	E	Exp	47	59
6	F	Ctrl	42	43
7	G	Ctrl	52	39
8	H	Exp	48	49

1. Select **Tables ▶ Stack** to open the dialog box shown in Figure 7.2.
2. The goal is to create a new column that represents the test administration as a categorical variable (pre- and post-) and stacks all of the scores into a single column. To do so, we select the two current columns (**Pre** and **Post**), and click **Stack Columns**.

   By default, JMP will name the new quantitative column **Data** and the categorical identifier (that is, the original column names) as **Label**. It's probably better to select more meaningful variable titles.
3. In the lower right of the dialog, specify a descriptive column title for the scores and for the experimental phase. Complete the dialog as shown in Figure 7.2 and click **OK**.

Figure 7.2: The Stack Dialog

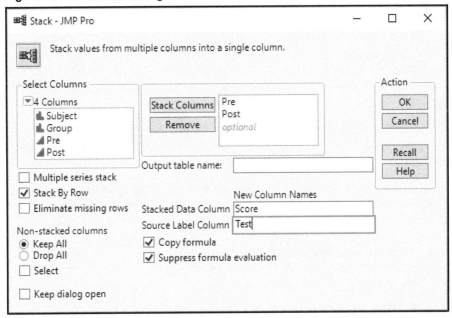

A new data table opens with the sixteen measurements represented in narrow (long) format. Your data table should now look like Figure 7.3.

Figure 7.3: The Experimental Data in Narrow, or Stacked, Format

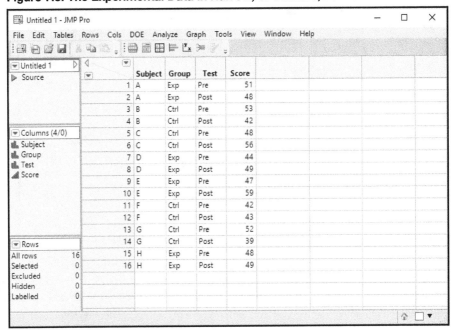

Unstacking Narrow Data

Suppose that the data were initially been recorded in the narrow format (similar to Table 7.2 and Figure 7.3), but we want to reorganize it in the wide format. To continue with the example and to reverse the process, we'll use the **Split** platform in the **Tables** menu. The dialog appears in Figure 7.4. Here we need to specify the column that identifies the basis for unstacking (in this case, the **Test** column) as well as the column containing the data that will populate the multiple new columns.

By default, JMP will drop all columns except those chosen in the split specification. For this example, it will make sense to retain the information about the subject and the experimental group.

1. So click the button next to **Keep All** under **Remaining columns** as shown in Figure 7.4. The dialog also provides the option to rename the new wide table.
2. Complete the dialog as shown and click **Create**.

Figure 7.4: The Split Dialog to Reshape a Narrow Table

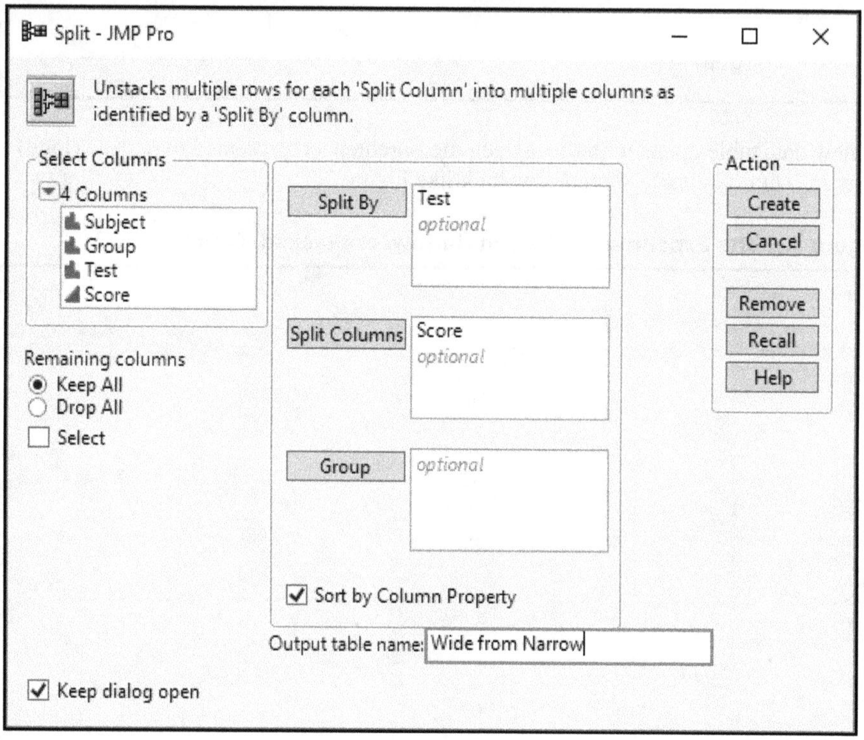

The results of this command appear in Figure 7.5. This nearly brings us full circle, with the one exception that JMP alphabetizes the order of the split columns. So, in contrast to Figure 7.1, the post-test data is now the third rather than the fourth column. You would need to note this in using the **Matched Pairs** platform for a pair-sample *t*-test or confidence interval.

> In JMP 13, the **Matched Pairs** platform is among the **Analyze ▶ Special Modeling** options. Veteran JMP users have previously found it in the **Analyze** menu.

Figure 7.5: The Experimental Data after Split to Wide Format

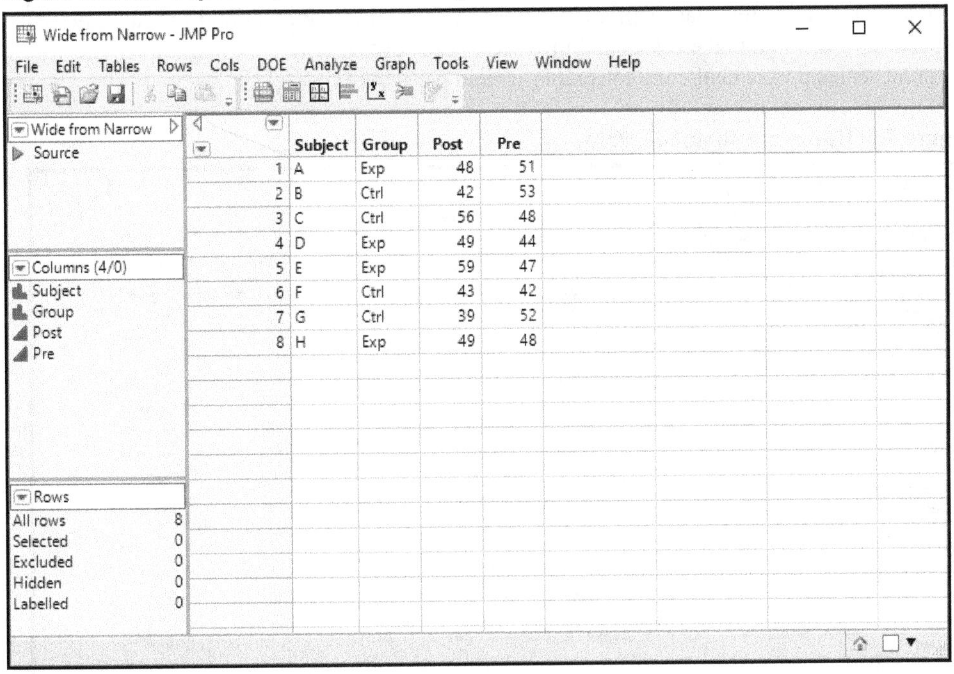

	Subject	Group	Post	Pre
1	A	Exp	48	51
2	B	Ctrl	42	53
3	C	Ctrl	56	48
4	D	Exp	49	44
5	E	Exp	59	47
6	F	Ctrl	43	42
7	G	Ctrl	39	52
8	H	Exp	49	48

Additional Examples

In practice, you would be unlikely to swap formats back and forth with a single set of data. Now that the **Stack** and **Split** commands are more familiar, let's continue with additional examples. remembering that often the analyst is working with existing data in a new context.

Stacking Wide Data

Wainer and Lysen (2009) report on a noteworthy historical example dating to 1951.

> Famous graphic designer Will Burtin (1908–1972) published a graphic display that was admired for the clarity and economy with which it showed the efficacy of three antibiotics on 16 different types of bacteria. The dependent variable was the minimum concentration of the drug required to prevent the growth of the bacteria in vitro—the minimum inhibitory concentration (MIC).

Interested readers can find the graphic by visiting the *American Scientist* web link to Wainer and Lysen's article. (See References.)

Burtin's intention was to clearly communicate the comparative efficacy of the three antibiotics, penicillin, streptomycin, and neomycin, in inhibiting the growth of each specific bacterium.

Burtin's data table is included in the JMP Sample Data Library as **Antibiotic MICs**, and is shown in Figure 7.6.

The 16 bacteria are in the rows, and the columns contain the genus and species for each, as well as three variables representing the MICs for each of the three antibiotics. Finally, the sixth column indicates whether the bacterium "the bacteria in question take up Gram stain or not. (The stain is named after its inventor, Hans Christian Gram [1853–1938])" (Wainer and Lysen 2009). An inspection of the data makes clear that the MICs span several orders of magnitude. The widely different scales pose a challenge for graphing, and we'll take that up shortly.

Figure 7.6: Burtin's Antibiotics Data

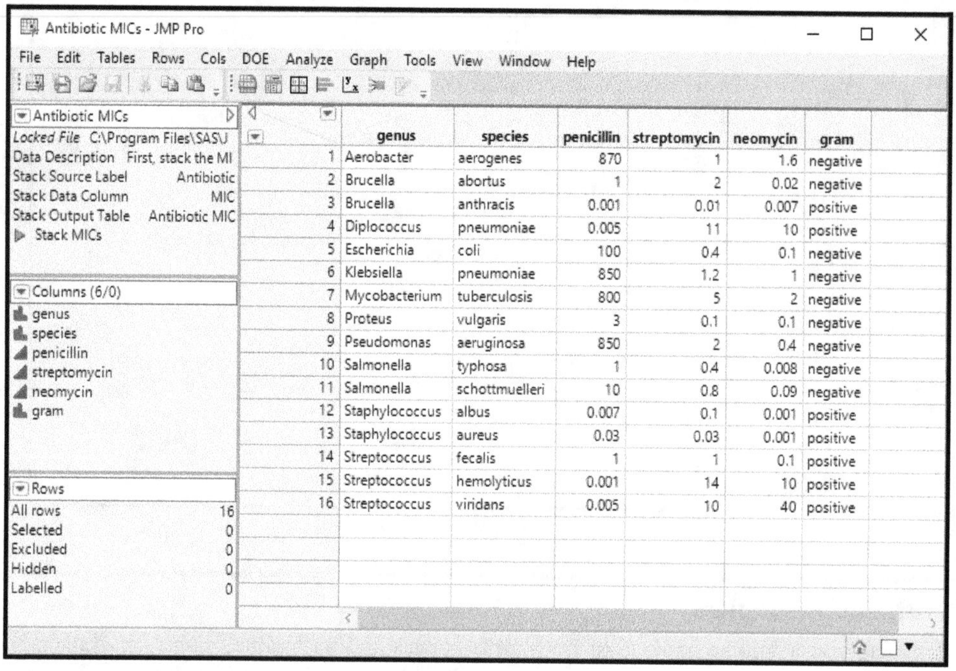

	genus	species	penicillin	streptomycin	neomycin	gram
1	Aerobacter	aerogenes	870	1	1.6	negative
2	Brucella	abortus	1	2	0.02	negative
3	Brucella	anthracis	0.001	0.01	0.007	positive
4	Diplococcus	pneumoniae	0.005	11	10	positive
5	Escherichia	coli	100	0.4	0.1	negative
6	Klebsiella	pneumoniae	850	1.2	1	negative
7	Mycobacterium	tuberculosis	800	5	2	negative
8	Proteus	vulgaris	3	0.1	0.1	negative
9	Pseudomonas	aeruginosa	850	2	0.4	negative
10	Salmonella	typhosa	1	0.4	0.008	negative
11	Salmonella	schottmuelleri	10	0.8	0.09	negative
12	Staphylococcus	albus	0.007	0.1	0.001	positive
13	Staphylococcus	aureus	0.03	0.03	0.001	positive
14	Streptococcus	fecalis	1	1	0.1	positive
15	Streptococcus	hemolyticus	0.001	14	10	positive
16	Streptococcus	viridans	0.005	10	40	positive

The wide layout served Burtin's purpose of comparing the quantity of each treatment to manage each species. However, for other types of analysis or visualization, a narrow (stacked) format is preferable. In fact, this data table includes a saved script as well as a usage note in the **Data Description** Table Variable. The description reads as follows:

```
First, stack the MIC columns (run the script).  Then, create a new
column and take the Log(MIC).

Background: In the fall of 1951 Burtin published a graph showing
the performance of the three most popular antibiotics on 16
different bacteria.

The response, the minimum inhibitory concentration (MIC),
represents the concentration of antibiotic required to prevent
growth in vitro. The covariate "gram" describes the reaction of the
bacteria to Gram staining.
```

The first paragraph above instructs us to create a new column equal to the logarithm of MIC. Why? Taking logs will accommodate the large differences in the magnitude of the MIC measurements. Logs and other functional transformations are discussed in Chapter 11. Rather than devote attention to the topic now, we'll defer it until that chapter.

Scripting for Reproducibility

Had we encountered the original raw data set and placed it into a JMP data table, our next step would be to use the **Stack** platform in the **Tables** menu, and specify that we want to stack three numeric columns (the three antibiotics), labeling the new source column as **Antibiotic,** the new data columns as **MIC**, and the new table as **Antibiotic MIC**.

The saved script does this for us with a single click. Before clicking, though, take note of the three table variables in the data table panel, shown in Figure 7.7. These variables are referenced in the saved script, so that the key inputs are visible to the user.

Figure 7.7: Table Variables as Inputs to a Script

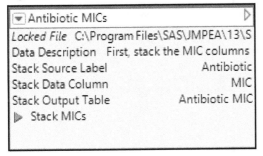

Hold the pointer over the green triangle, right-click. and choose **Edit** to open the script before running it. Figure 7.8 shows the JSL code that stacks the data table. Rather than explicitly using the new column and table names in the code, the script refers to expressions using the syntax `Expr(`.

Figure 7.8: JSL Code to Stack the Antibiotics MIC Data

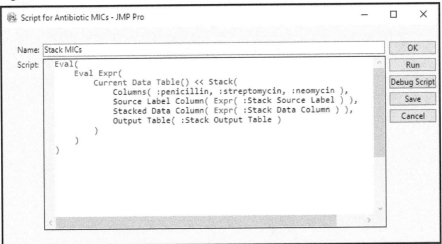

1. At this point, click **Cancel** and then click the green triangle next to **Stack MICs**. This opens a new data table shown in part in Figure 7.9. The new data table now has three rows per bacterium and two new columns labeled **Antibiotic** and **MIC**. All of the data elements from the original table have been retained, but are in a new arrangement.
2. As a challenge to the reader, see whether you can use the **Tables ▶ Stack** platform and the instructions within the script (Figure 7.8) to achieve the same end.

Figure 7.9: Antibiotics Data in Stacked (Long) Format

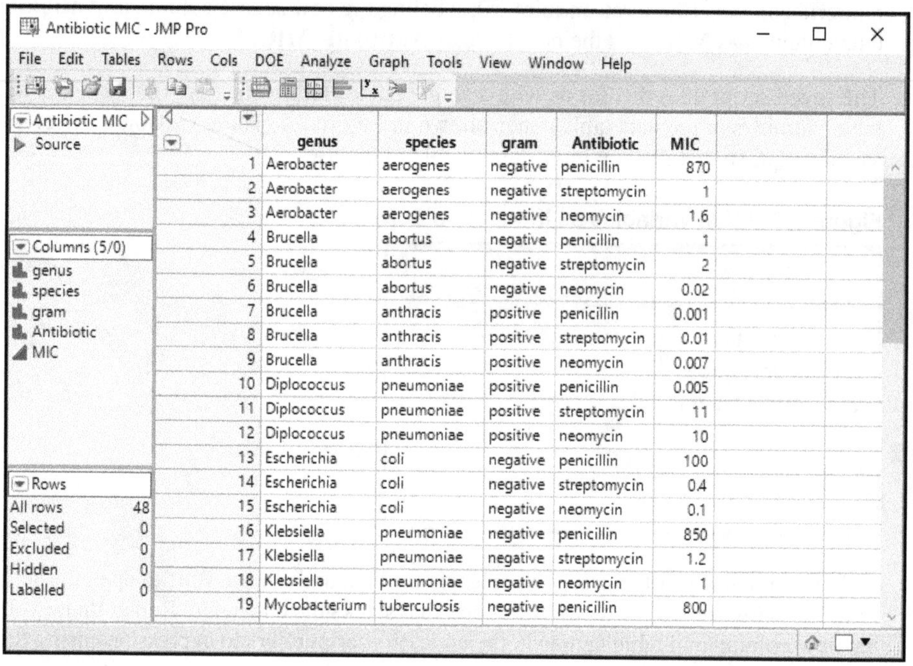

	genus	species	gram	Antibiotic	MIC
1	Aerobacter	aerogenes	negative	penicillin	870
2	Aerobacter	aerogenes	negative	streptomycin	1
3	Aerobacter	aerogenes	negative	neomycin	1.6
4	Brucella	abortus	negative	penicillin	1
5	Brucella	abortus	negative	streptomycin	2
6	Brucella	abortus	negative	neomycin	0.02
7	Brucella	anthracis	positive	penicillin	0.001
8	Brucella	anthracis	positive	streptomycin	0.01
9	Brucella	anthracis	positive	neomycin	0.007
10	Diplococcus	pneumoniae	positive	penicillin	0.005
11	Diplococcus	pneumoniae	positive	streptomycin	11
12	Diplococcus	pneumoniae	positive	neomycin	10
13	Escherichia	coli	negative	penicillin	100
14	Escherichia	coli	negative	streptomycin	0.4
15	Escherichia	coli	negative	neomycin	0.1
16	Klebsiella	pneumoniae	negative	penicillin	850
17	Klebsiella	pneumoniae	negative	streptomycin	1.2
18	Klebsiella	pneumoniae	negative	neomycin	1
19	Mycobacterium	tuberculosis	negative	penicillin	800

Splitting Long Data

Our next example involves data about the comparative market shares of different providers of operating systems for smartphones in the United States for the years 2006-2011. Here again we have longitudinal data, and also once again a data table is included in the JMP Sample Data Library. Open the file titled **Smartphone OS**. (See Figure 7.10.)

Figure 7.10: Smartphone OS Data Table

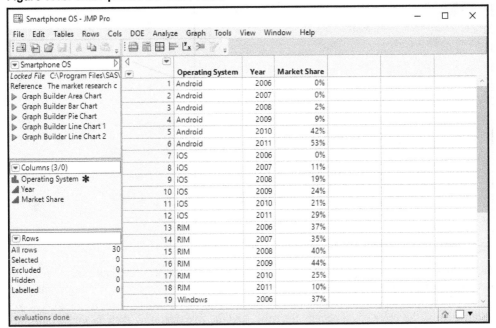

For each observed year, we have the market share for four different firms and an aggregate group, as follows: Apple iOS, Google Android, Microsoft Windows, Research in Motion (RIM)/Blackberry, and Other. The arrangement is long, with **Operating System** and **Year** as identifier columns, and **Market Share** as a data column.

1. In the **Columns** pane, notice the asterisk next to **Operating System**.
2. Right-click and select **Column Info.** (See Figure 7.11.) Under **Column Properties**, you should see **Value Ordering** and **Value Colors**. In this particular data table, the creator of the table specified the order in which the operating systems should be listed in analyses and the colors used to represent the different systems.

Figure 7.11: Column Properties

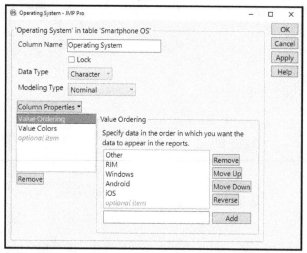

Depending on the modeling plan, we might want to reshape this as wide data, and we need to decide which of the two identifier columns we want to recast as variables. We might imagine having five rows—one for each operating system—and six data columns corresponding to years. Alternatively, the years could occupy the rows, and the five operating systems could make up the columns. The decision depends on the plan for analysis. For this illustration, we'll use the latter approach, so that each row ultimately will contain the market shares for a given year.

In short, we want a table with six rows and six columns. The first column will be the years, followed by five more columns for each operating system. The values in the cells of the table will be the market shares.

1. Select **Tables ▶ Split**.
2. Complete the dialog as shown in Figure 7.11, being sure to check **Keep dialog open** in the lower left. Notice that when that check box is ticked, the **OK** button under **Action** in the upper right is relabeled **Create**.

Figure 7.12: The Split Dialog

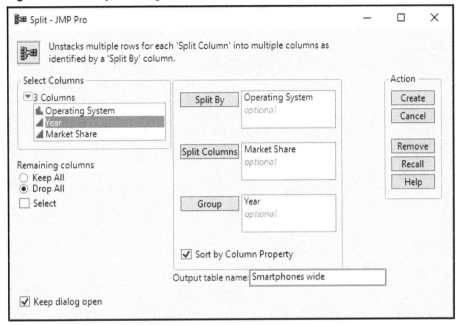

3. Click **Create**. You will now see a new data table, reproduced in Figure 7.13. Note that by default there is a check in the box labeled **Sort by Column Property**. When there is a previously defined value ordering in the data table, this option tells JMP to use the value ordering that was set as a column property rather than the default alphabetical ordering.

Figure 7.13: The Smartphone OS Data in Wide Format

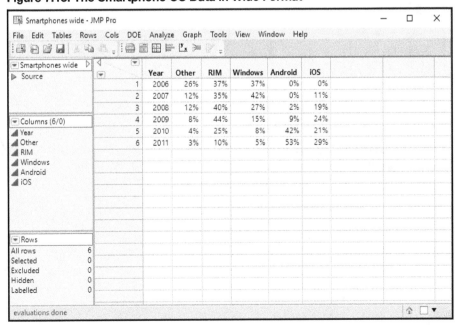

Suppose we wanted the layout with operating systems in the rows and years across the columns. Glancing back at the still-open **Split** dialog, we would simply swap the Split by and Group columns. Readers are encouraged to return to the dialog and try making the change.

Transposing Rows and Columns

Alternatively, with the wide format shown in Figure 7.12, we might *transpose* the data matrix using the **Tables ▶ Transpose** platform to interchange the rows and columns in the table. In the dialog (Figure 7.14), place the five operating system columns into the **Transpose Columns** drop zone, select **Year** as **Label**, name the new output table as **OS by Year**, and type **Operating System** as the **Label column name**. Click **OK**.

Figure 7.14: Transpose Dialog

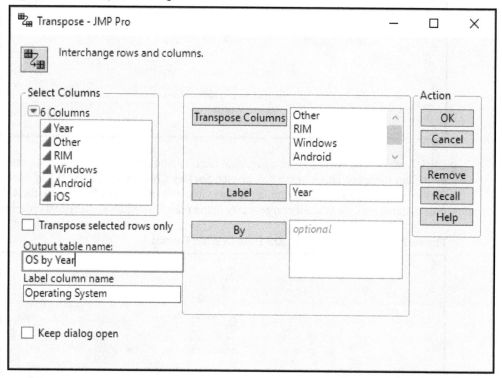

Figure 7.15 shows the new output table. There is a column called **Operating System** to identify each row, and the market share percentages appear in the yearly columns.

Figure 7.15: Transposed Operating System Market Shares

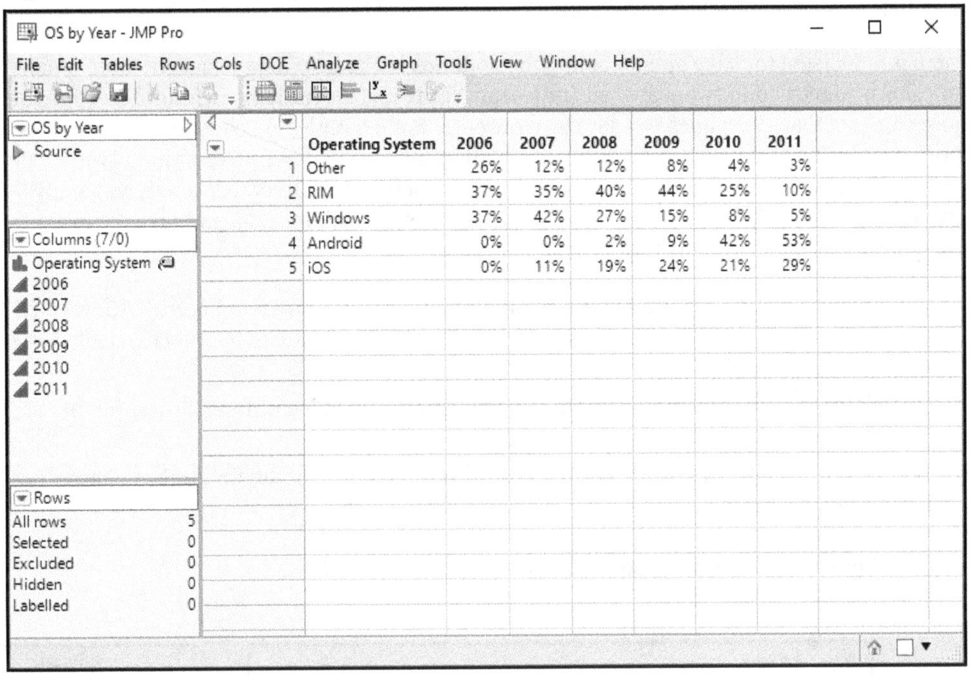
OS by Year - JMP Pro
File Edit Tables Rows Cols DOE Analyze Graph Tools View Window Help
OS by Year
Source
Columns (7/0)
Operating System
2006
2007
2008
2009
2010
2011
Rows
All rows 5
Selected 0
Excluded 0
Hidden 0
Labelled 0
Operating System 2006 2007 2008 2009 2010 2011
1 Other 26% 12% 12% 8% 4% 3%
2 RIM 37% 35% 40% 44% 25% 10%
3 Windows 37% 42% 27% 15% 8% 5%
4 Android 0% 0% 2% 9% 42% 53%
5 iOS 0% 11% 19% 24% 21% 29%

Reshaping the WDI Data

In Chapter 4, we encountered the World Development Indicators (WDI) data from the World Bank. In that chapter, we downloaded the raw data and saved them as a JMP data table called **WDI_Data**. The data table has 60 columns. The first four are nominal variables labeling the nation and the indicator (measurement), and the remaining 56 are continuous annual data. In the raw data download, there are 248 countries and groups of countries (for example, "Arab World"). For each country and group of countries, there are 1,420 different series or measures. So, the data table consists of the 60 columns and 352,160 rows. Each row contains a unique pairing of a country and a series (248 country groups x 1,420 series equals 352,160). This is a wide format where each of the 1,420 World Development Indicators—the variables that we might wish to model—appears in 248 non-contiguous rows.

For example, the first data series is named "2005 PPP conversion factor, GDP (LCU per international $)." The PPP (purchasing power parity) conversion factor allows for comparing monetary measures across national currencies. Thus, it is an important factor in interpreting international financial information. The WDI data set contains this figure for each country annually, but after it appears in row 1 for Afghanistan, we do not see it again until row 1,421 for Albania.

As indicated in Chapter 4, a few moments scrolling through this rather large data table reveal numerous data preparation tasks ahead. For the purposes of this chapter, we first take on the reshaping tasks so that the data series appear in columns rather than in rows. This is not a simple matter of transposing rows and columns, because we want to have each country appear in a row 56 times for each of the 56 years and have each variable occupy a separate column. We also want to retain the two columns with the names and codes of the countries. We'll accomplish this in two steps. First, we'll stack all of the data so that we treat year as a variable. Then we'll split out all of the different series into separate columns.

There is a decision to make about naming the columns in the new long (stacked) table. Each indicator is identified in two ways: an **Indicator Name** and an **Indicator Code**. The codes are more compact than the names, yet far more obscure. For example, the code for purchasing power parity conversion factors is PA.NUS.PPP.05, which is hardly a descriptive or intuitively resonant title. On the other hand, the full names will make for hard-to-decipher axis labels and analysis reports. For the sake of clarity, we'll use the full names to stack the data and save the task of renaming a subset of columns for the next chapter.

1. Once again, we use the **Tables ▶ Stack** platform. As shown in Figure 7.15, we will stack all of the year columns from **1960** through **2015** into an output table called **WDI Long**.
2. Name the data column Data and the new label column Year. In the lower left of the dialog.
3. Check **Select** under **Non-stacked columns**, and highlight the first three columns before clicking **Create**.

Figure 7.16: Specifications for the Stack Command

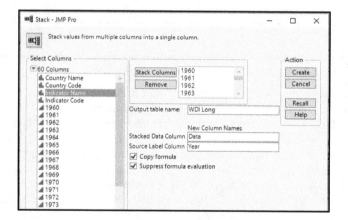
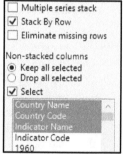

This will take a few moments, and a window will open showing the progress of the operation. The new table should have 5 columns and 19,720,960 rows. For each country, every series appears 56 times now alongside the new Year column, which runs from 1960 through 2015 repeatedly. From this table, we can split the data into a format in which most of the columns are different series. But first, we should make one adjustment.

The **Columns** panel in the **WDI Long** data table indicates that the new **Year** variable is nominal, as indicated by the red bars: ▆▆▆Year. Going forward, it will be wise to treat year as numeric.

4. Select the **Year** column and select **Cols ▶ Column Info**. Change the **Data Type to Numeric**, the **Modeling Type to Continuous**, and click **OK**.

5. Select **Tables ▶ Split**, and complete the dialog as shown in Figure 7.16. The indicator names will become the additional new columns; the cells will contain the numeric data; and observations will be grouped by country and year. This operation will also take a few moments. But when it's complete, you will see the final data table with 1,423 columns and 13,888 rows (248 countries x 56 years). The resulting data table is shown, in part, in Figure 17.17.

Figure 7.17: The Split Dialog

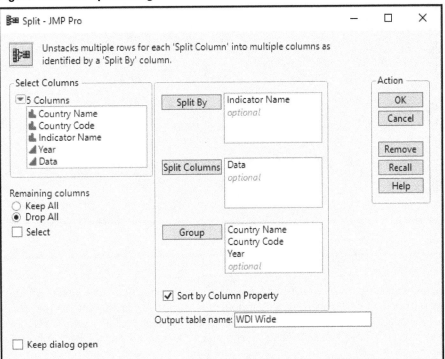

Figure 7.18: The WDI Data Reshaped

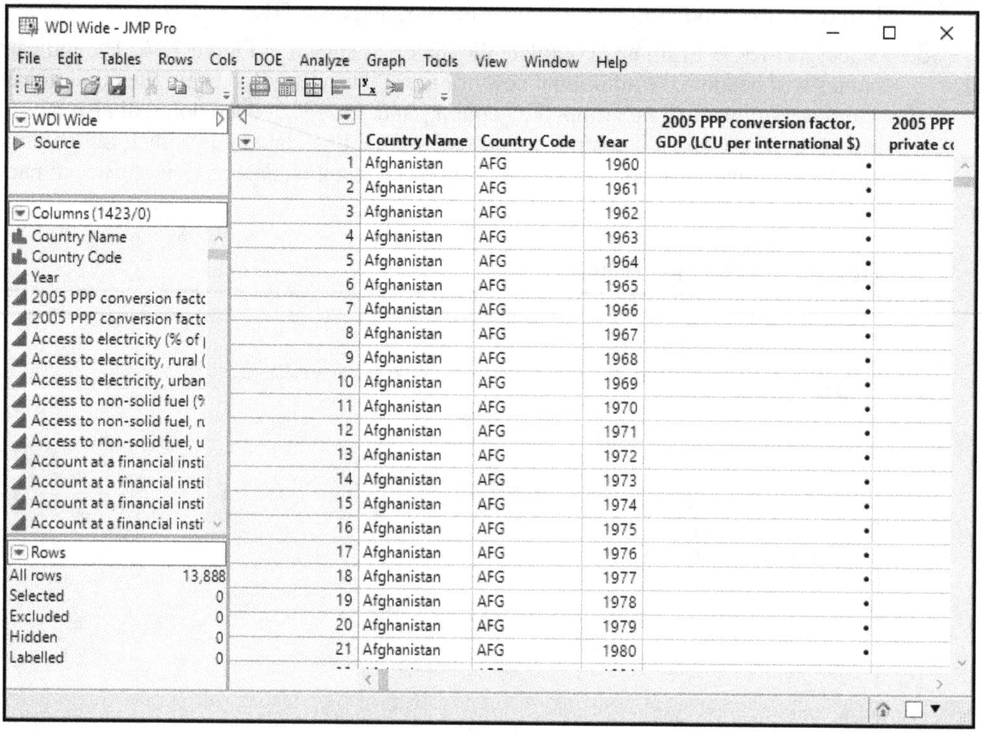

Conclusion

This chapter has introduced three common tasks in reshaping a data table along with the corresponding commands in the JMP **Tables** menu. Reshaping data is one invaluable stage in the process of preparing a data table for further analysis and modeling. In the next chapter, we take up additional **Tables** menu commands that enable us to combine several tables and to create subsets based on columns and/or rows.

References

Codd, E.F. 1990. *The Relational Model for Database Management: Version 2*. Reading, MA: Addison-Wesley.

Grace-Martin, Karen. 2015. "The Wide and Long Data Format for Repeated Measures Data." Available at http://www.theanalysisfactor.com/wide-and-long-data/.

SAS Institute Inc. 2016a. *JMP 13 Fitting Linear Models*. Cary, NC: SAS Institute Inc.

SAS Institute Inc. 2016b. "Sample 30584: Analyzing Repeated Measures in JMP Software." Available at http://www.jmp.com/support/notes/30/584.html.

Wainer, Howard, and Shaun Lysen. 2009. "That's Funny…A window on data can be a window on discovery." *American Scientist* (July/August). Available at http://www.americanscientist.org/issues/pub/2009/4/thats-funny.

Wickham, Hadley. 2014. "Tidy Data." *Journal of Statistical Software*. August (59:10).

Wickham, Hadley. 2016. "Package 'reshape2'." The Comprehensive R Archive Network. Available at https://cran.r-project.org/web/packages/reshape2/reshape2.pdf.

World Bank. 2016. "World Development Indicators." Available at http://data.worldbank.org/data-catalog/world-development-indicators/.

Chapter 8: Joining, Subsetting, and Filtering

Introduction

In Chapter 5, we noted the likelihood that any particular project might require only some of the data stored in a particular table or it might require data from different tables. In that chapter, we saw how JMP can facilitate database operations as part of extracting and importing data into one or more JMP data tables.

Of course, a given project might demand similar operations on a user's collection of JMP data tables. An analyst might want to join multiple JMP tables to carry out an analysis. Alternatively, an analysis might run intolerably slowly on a table of all available data, with little substantive benefit. Hence, the analyst might want to work with a subset.

Also, for time series modeling or machine learning tasks, it might be advisable to create separate subsamples for model development and model validation. For some projects, an analyst might need to create a separate data table containing the subset. But, for others, it might be sufficient to temporarily filter out certain rows. JMP offers several automated approaches to partitioning large data sets for validation, and this chapter illustrates one method.

In these and other cases, the analyst might need to join JMP tables on a shared column or to subset by column, rows, or both. This chapter works through several examples involving such operations, all of which can be accomplished via platforms and tools in the JMP **Tables** and **Rows** menus. Specifically, this chapter focuses on **Join**, **Subset**, **Concatenate** and **Query Builder** for tables as well as Row Selection, Hide/Exclude, and Data Filter in the Rows menu.

Combining Data from Multiple Tables with Join

Our first example is inspired by the MovieLens.org website, which provides personalized movie recommendations to users. Here's the description from their website:

> MovieLens is run by GroupLens, a research lab at the University of Minnesota. By using MovieLens, you will help GroupLens develop new experimental tools and interfaces for data exploration and recommendation. MovieLens is non-commercial, and free of advertisements. (MovieLens 2016)

GroupLens maintains several data collections. In this example, we use a set of four tables of user-supplied ratings for a collection of movies. Readers should note that the researchers periodically update this set of data, with the most recent version accessible at https://grouplens.org/datasets/movielens/latest/ (GroupLens 2016).

The files are distributed in a single zipped archive of *.csv files, which I imported using the techniques illustrated in Chapter 4. In this section, the four filenames are preserved, except with the suffix *.jmp replacing *.csv. Here is a description of the contents of the four files, abstracted from the MovieLens site:

> This data set (**ml-latest-small**) describes 5-star rating and free-text tagging activity from MovieLens, a movie recommendation service. It contains 105339 ratings and 6138 tag applications across 10329 movies. These data were created by 668 users between April 03, 1996 and January 09, 2016. This data set was generated on January 11, 2016.
>
> Users were selected at random for inclusion. All selected users had rated at least 20 movies. No demographic information is included. Each user is represented by an ID, and no other information is provided.
>
> The data are contained in four files: links.csv, movies.csv, ratings.csv, and tags.csv.
>
> Ratings are made on a 5-star scale, with half-star increments (0.5 stars - 5.0 stars).
>
> Timestamps represent seconds since midnight Coordinated Universal Time (UTC) of January 1, 1970.

Suppose an analyst wants to examine the distribution of ratings for all movies across all raters. For that purpose, all of the relevant data is in the tables **movies** and **ratings**. Figure 8.1 shows the Columns panels for the two data tables, with **movies** to the left.

Figure 8.1: Columns in Movies.jmp and Ratings.jmp

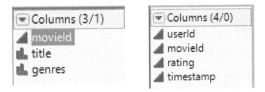

The movies table contains a simple integer code identifying the film, along with the film title and a multiple-response column that lists the genres by which the film could be classified, such as "Comedy|Romance" or "Drama|Sci-Fi." For this analysis, we'll want just the first two columns.

The ratings table has four columns that identify the user who rated a film, the film identification code, a numeric rating, and the timestamp described above. In this simple example, **userid**, **movieid**, and **rating** are the columns required.

Both tables have a column named **movieid** that was imported as continuous. Given its role as an identifier, it makes more sense to change the modeling type to nominal. So we should click on the blue triangle next to it in the **Columns** panel and make the change. Similarly, the identifier for the user should also be nominal.

At this point, it's worth noting that for other analyses, we'd need to make other modifications to the raw data that was imported. The film titles all include the year of release within parentheses, such as "Toy Story (1995)". Depending on the goals of the analysis, it might be preferable to split titles into two columns. Also, we would probably change the modeling type of genres to **Multiple Response**, and do some further work to render the timestamp into a recognizable date and time. Those changes will be demonstrated in Chapters 9 and 11.

For present purposes, we want to join the two tables on their shared column, creating a new output table. Because the primary interest is in the ratings, we'll start with that table.

1. Select **ratings** as the active table.
2. Select **Tables ▶ Join** to open the dialog shown in Figure 8.2. In this dialog, we specify the tables to be joined, the columns from each table to preserve in the resulting output table, and set options about the treatment of formulas. JMP provides three methods for matching tables, but we'll only use the default method here of joining on a column.

Figure 8.2: The Join Dialog

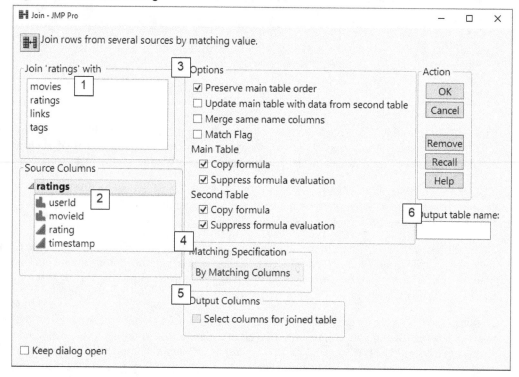

3.   In the panel labeled **Join 'ratings' with** (numbered 1 in the figure), select **movies**. This will open a second panel under **Source Columns** (2).

4.   Under **Options** (3), we'll use all of the defaults except that we want to check **Merge same name columns**. This is because our two source files share a key field with a consistent name in the two tables. Checking the box automatically checks the **Match Flag** option to create a new nominal column in the output table that indicates the origin of each row of data. Uncheck this box to omit that column.

5.   Once a second source table is selected, the **Matching Specifications** panel expands to provide options for matching. The default is an inner join based on the shared column. Highlight **movieid** under *both* Source Columns tables, and click **Match** (Figure 8.3). Though we won't make any further changes here, note that we can use these options when the shared key field has a different name in the two tables. Also, by checking one or both **Include non-matches** boxes we could perform a left-, right-, or full-outer join. Sometimes, we want all rows from the main table, whether there are matching rows in the secondary table. For other investigations, we could want all rows from the second table, even if they don't match an entry in the first. As you check and uncheck the boxes related to non-matching rows, notice that the Venn diagram icon changes in response to indicate how the resulting join will occur.

     This section also includes an option to **Drop Multiples** in either or both tables. Though we don't need this option here, it can be useful if there are multiple rows in either table with the same matching key column. Checking the boxes next to **Drop Multiples** will indicate that you want to write only the first match found into the new joined table. Checking both boxes under both tables just keeps the first match found. Checking the box under one but not the other will keep the first match in the table checked, and join it to all matches in the other table. For further documentation, see the "Join Data Tables" entry in Chapter 6 of *Using JMP*.

6. Under **Output Columns**, check **Select columns** for joined table. Then in the **Source Columns** boxes, highlight only the columns that you want for the new result table, and click **Select** as showing in Figure 8.3.

Figure 8.3: Choosing the Columns to Include in the Joined Table

7. Finally, type a descriptive name for the resulting table, such as **movieratings**, into the **Output** table name box (6 in Figure 8.2), and click **OK**.

The first 20 rows of the output table are showing in Figure 8.4. There are 105,339 rows—the same as the total number of ratings. The first 232 rows are ratings of Toy Story (1995) by different individuals, followed by 92 ratings of Jumanji (1995), and so on. The data table contains only the columns that we chose.

Figure 8.4: The Results of the Join Operation

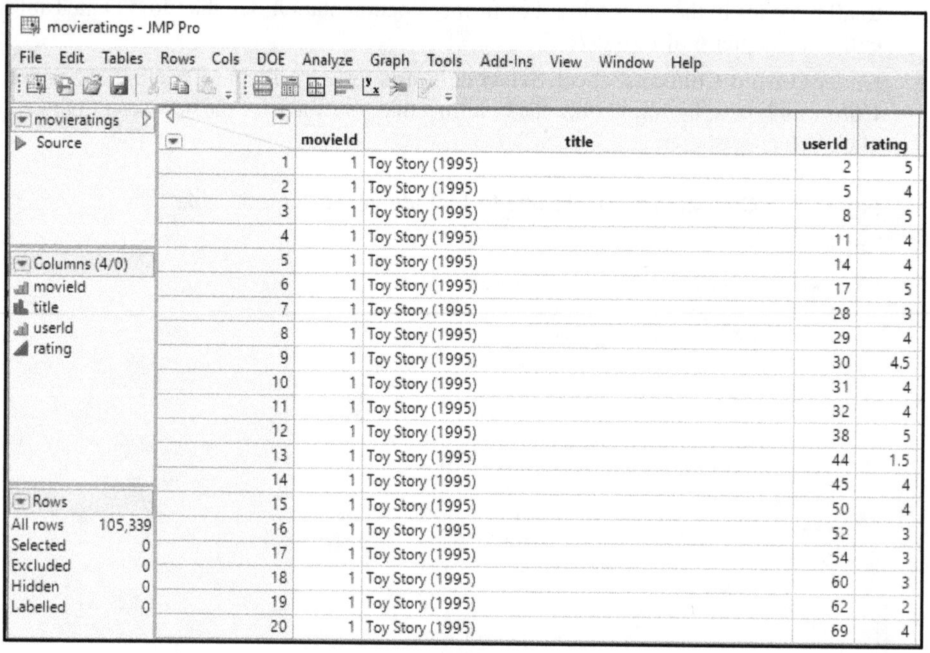

	movieId	title	userId	rating
1	1	Toy Story (1995)	2	5
2	1	Toy Story (1995)	5	4
3	1	Toy Story (1995)	8	5
4	1	Toy Story (1995)	11	4
5	1	Toy Story (1995)	14	4
6	1	Toy Story (1995)	17	5
7	1	Toy Story (1995)	28	3
8	1	Toy Story (1995)	29	4
9	1	Toy Story (1995)	30	4.5
10	1	Toy Story (1995)	31	4
11	1	Toy Story (1995)	32	4
12	1	Toy Story (1995)	38	5
13	1	Toy Story (1995)	44	1.5
14	1	Toy Story (1995)	45	4
15	1	Toy Story (1995)	50	4
16	1	Toy Story (1995)	52	3
17	1	Toy Story (1995)	54	3
18	1	Toy Story (1995)	60	3
19	1	Toy Story (1995)	62	2
20	1	Toy Story (1995)	69	4

Saving Memory with a Virtual Join

One disadvantage of joining tables is that we now have data values from the source tables actually replicated in the output table. With large data sets or particularly large data values (for example, full-text or images), such duplication can be inefficient. JMP provides the alternative of a *virtual join* to temporarily access data from auxiliary tables by setting column properties to identify linked variables.

To establish a virtual join, we need at least one table in which the linking column consists entirely of unique, non-repeated values. This will be the auxiliary or *referenced* table. Here, that's the **movies** table. We first need to set two new column properties as follows:

1. In the secondary table (**movies**), open the **Column Info** for the **movieid** column. Click **Column Properties**, and select **LinkID** near the bottom of the list. Tagging this column as the LinkID basically identifies it as a primary key that can be referenced from another table.
2. In the ratings table, open the **Column Info** for **movieid.** Click **Column Properties**, and select **Link Reference**. This opens a new panel asking for a **Reference Table**. Click **Select Table** and choose **movies.jmp** from your directory.

In Figure 8.5, we see the two data tables after establishing the Link ID and Link Reference properties. In the referenced table, there is a gold key next to the linked column name. In the main table, the same column is flagged as the *referring column*. Below the list of available columns are additional accessible columns from the referenced table.

Figure 8.5: Two Virtually Joined Tables

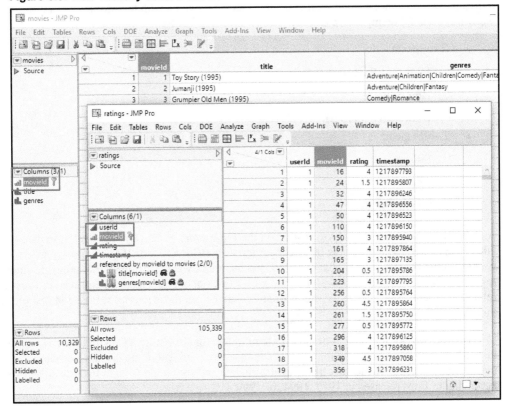

Now any analysis using the ratings table can use its four columns as well as the two additional columns from the movies table. So, though this is not a literal join resulting in a new table, it virtually joins the data from the two tables.

Why and How to Select a Subset

As with so many other steps in data management and preparation, the goals of an analysis drive decisions about analyzing just a portion of the available data. In situations where one or more subsets are desirable, you must also decide if the subsetting should be a temporary filtering or if it would be better to create and save new data tables for the subsets.

For example, you might want to temporarily partition a large data set into training and testing sets, randomly assigning rows to one subsample or another. For cross-validation, we might work with numerous subsets.

In other projects, it might be more functional to create and save a new data table consisting of only selected rows and columns from a larger table. Or, we might want to aggregate values in a table by one or more columns.

We'll look at illustrative examples of such inquiries. JMP has several approaches that can be used individually or in concert to subset a data table. As we move on with the Movie Lens ratings data, we will use the following:

- **Rows** menu
  - **Hide and Exclude**
  - **Row Selection**
  - **Data Filter (global)**
- **Tables** menu
  - **Subset**
  - **Concatenate**
  - **Query Builder**

As a first example, consider a blog posting on the *Entertainment Weekly* website (Nashawaty 2009) that asked "Which was the best year for movies: 1977, 1994, or 1999?" In our joined table, we have some data relevant to that question, and we might want to compare the ratings for movies released in those three years. Recall that the movie titles in our data table all end with the year of release in parentheses. Hence, we might want to analyze just those movies released in one of the three years in question.

This research question is typical of a use-case rationale for subsetting a data table: Of the more than 100,000 rows of data available, only some rows are pertinent to the question at hand. It makes sense to set aside the bulk of the available data and subset the table to include only rows pertaining to films released in 1977, 1994, or 1999.

Let's look at two approaches to this task, one that temporarily ignores all years other than these three and one that creates a new data table of movies just from the three years under consideration. Prior to following either approach, it will be helpful to take the time to create a new column that contains the year of release. We do this by using some character functions to isolate the release year from the title column.

A Brief Detour: Creating a New Column from an Existing Column

The **title** column includes the data we want, but it's not tidy. An entry like "Toy Story (1995)" contains two attributes of the film within one column. For data management and analysis, it is preferable to separate the two data elements into two columns. In JMP, this requires a simple formula. Fortunately, every title follows the same format of *"title (yyyy)"*. To create a column of release years, we want to extract four digits from the last five characters of the title field. Because the years are discrete and we intend to use individual years for filtering, we'll make the new column ordinal.

Let's do this operation on the full table of **movieratings** rather than the subset we just created, so that it will be possible to use the new release year data in other ways as well.

1. Make **movieratings** the active table.
2. Select **Cols ▶ New Column** to open the **New Column** dialog (Figure 8.6), and give the new column a descriptive name like **year**.

Figure 8.6: Starting to Create a New Column

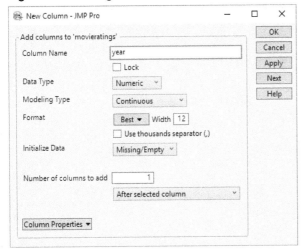

Though **title** is a character variable, we can use the default data and modeling types—numeric and continuous—in creating this new column. Next, we need to create the formula to find the year within the title. Before building the formula, let's briefly consider the logic required and introduce the character functions that will do the job.

If a particular movie title is k characters long, then the year is found starting at the $(k-4)^{th}$ position. For example, the title "Toy Story (1995)" is 16 characters in length. The 16th character is the close parenthesis. Positions 12–15 are the digits 1995. So, in general, we want to pull out characters $(k-4)$ through $(k-1)$. Below, we'll use the JMP **length** function to find the length of each title string, and the substring (**substr**) function to isolate the four characters for the year.

In other words, if k is the number of characters in the title of a movie, the year is a substring of the title that begins at position $(k-4)$ and is 4 characters long. With this logic in mind, do the following steps. There are several ways to create the formula, and, in this illustration, we'll use the Formula Editor to build it one element at a time.

1. At the bottom left of the dialog, click **Column Properties**, and choose **Formula**.
2. As shown in Figure 8.7, click the disclosure triangle next to **Character** in the list of function groups, and highlight title in the **Columns** panel. This places *title* in the center of the formula enclosed in a blue rectangle.

Figure 8.7: Character Functions in the Formula Editor

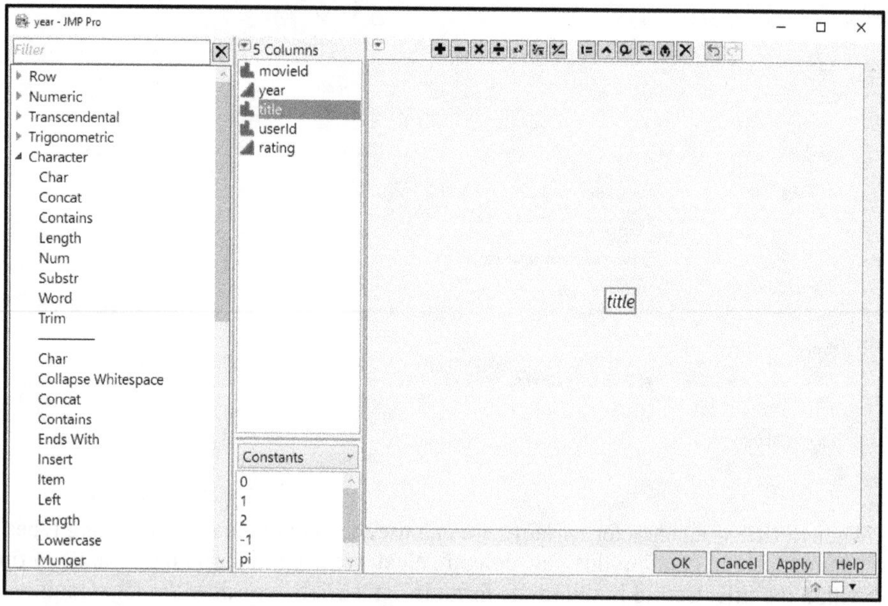

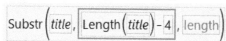

3. In the list of **Character** functions, click **Substr**. The formula will now look like Figure 8.8. The substring function requires three arguments: the column containing the source string, the starting position of the substring, and the length of the substring.

Figure 8.8: Substringing the Title Column

$$\text{Substr}\left(\textit{title}, \textit{start}, \textit{length}\right)$$

4. Within the formula, click the term start (second argument). This is where we want to say "start at position $(k - 4)$." We need the Length function to represent k, so click Length in the list of character functions.

5. In the formula, you will now see the Length function awaiting a column name. Click **title** in the list of 5 columns.

6. In the formula box, double-click **Length**$(\textit{title})$ in the blue rectangle. This opens the formula editor, showing the expression Length(:title). Move the cursor to the right end of the expression, click, type – 4, and press **Enter**. The formula now looks like Figure 8.9.

Figure 8.9: Defining the Starting Position of the Substring

$$\text{Substr}\left(\textit{title}, \text{Length}\left(\textit{title}\right) - 4, \textit{length}\right)$$

7. Click length at the end of the expression, type 4, and click **OK**.

   This returns us to the **New Column** dialog, showing the formula at the bottom of the dialog. Click **OK**, and find the new **year** column in the data table.

8. Save the data table as **movieratings with year**.

Now we can easily identify movies released in any of the three years of interest. Next, we'll use the Data Filter to choose the rows we want for analysis. After we select rows, we can either Include and Show only the selected rows, or we can subset the data table, building a new table containing only movies from the three years 1977, 1994, and 1999.

Row Filters: Global and Local

The JMP **Data Filter** has two modalities. It can be applied to a data table during a session and affect all analyses for as long as it is active, or it can be invoked within a particular analysis platform to affect only that one report. The former is known as *global* filtering, and the latter as *local* filtering. In this section, we'll demonstrate both modes.

Global Filter

Recall that the years of interest are 1977, 1994, and 1999. We'll filter down to select just those rows, and use **Graph Builder** to explore the ratings, starting with a global filter.

1. First, to make sure that prior row selections are clear, select **Rows ▶ Clear Row States**.
2. Now select **Rows ▶ Data Filter**.
3. In the **Add Filter Columns** panel, highlight **year** and click **Add**.
4. At the top of the dialog, click **Show** and **Include**.

> **Useful Tip**: For filtering based on ordinal or nominal variables, Data Filter's default selector is a drop-down list from which you can select items individually, in a contiguous group, or with a multiple selection using the **Ctrl** key.

> In a case like this, where you want to choose disparate years, you can change the selector to a list with check boxes, as shown in this step.

5. Within the filter, click the red triangle next to **year** and notice the four Display Options.
6. Scroll through the list of years and highlight **1977**, **1994**, and **1999** (shown in part in Figure 8.10). At the top of the dialog, notice that 11403 rows match the specification just as before.

Figure 8.10: Selecting Criteria for the Global Data Filter

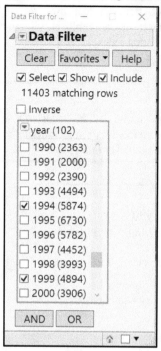

7. The dialog remains open, so minimize it or just shift it out of the way. Now return to the data table, and select **Graph ▶ Graph Builder**.

8. Place **rating** on the **Y** axis and **year** on the **X** axis.

9. Right-click and **Add** a **Box Plot**.

Figure 8.11 shows one version of the resulting graph. Our focus is on data wrangling and preparation. So we resist the temptation to draw preliminary inferences about the best years in movie making.

Figure 8.11: Graph Builder Exploration of Filtered Data

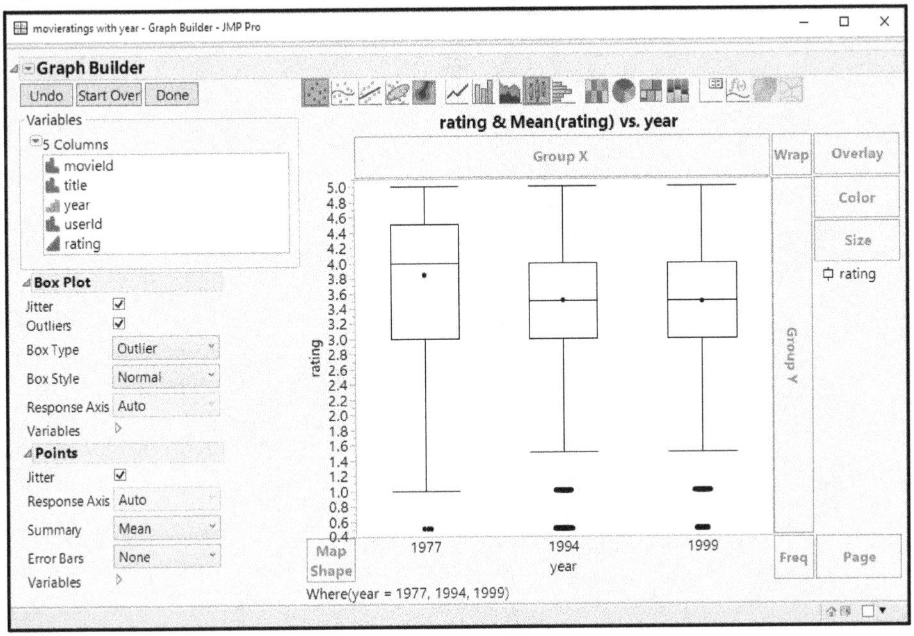

Local Filter

We also have the option of applying a filter within an analysis platform to apply just to the current analysis, rather than to the entire data table.

- Reopen the **Data Filter** dialog, click **Clear**, and then close the window.
- Return to the **Graph Builder** window. Note that now the graph has changed dramatically as all years are displayed.
- At the upper left of the window, click the red triangle next to **Graph Builder**, and select **Local Data Filter**.
- This opens a new panel in **Graph Builder** (Figure 8.12). This operates exactly the same way as the global filter. Choose the three years and see the graph change.

Figure 8.12: The Local Data Filter

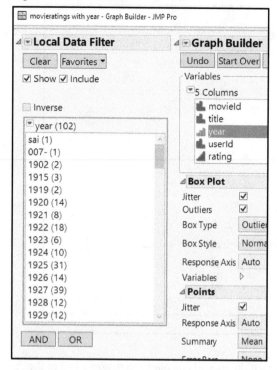

A More Durable Subset

The approach just demonstrated leaves all of the original data untouched, although some of it is suppressed for the time being. The last two steps above might strike some as roundabout, but, in any event, they leave a situation in which we'd be ignoring the vast number of rows in the data table.

To peel off a subset into a fresh new data table, we could do the following instead of inverting the row selection and hiding/excluding the non-selected rows.

1. Return to the movieratings with year data table. Notice that no rows are selected, because the filtering we just did applied only to the Graph Builder. Go back and use the global data filter to choose the years 1977, 1994, and 1999.

2. Select **Tables ▶ Subset** to open the **Subset** dialog (Figure 8.13). There are several options available, but in this instance we want to use the defaults. Click **OK**.

Figure 8.13: The Subset Dialog

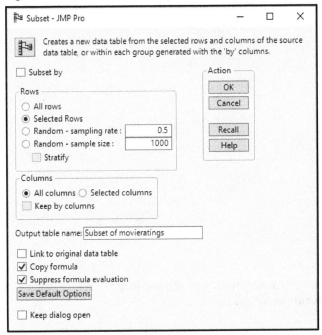

The result is a new data table (with a source script embedded as a table variable) containing only movies from one of the three years.

Combining Rows with Concatenate

Thus far, we've looked exclusively at situations where we need to combine data columnwise. But recall the financial institution example from Chapter 5. The analytic objective was to build a model to identify clients most likely to be interested in an insurance investment product. After querying several SAS data tables, we created two JMP data tables. The first contained data about customers who had already purchased an insurance product, and the second was a random sample of customers who had not. The first table had 15 columns and the second had 11. Ten of the column names are identical though the columns occur in different sequences. Naturally, the first table also included specific data about the insurance product that the client had invested in. Figure 8.15 shows the column panels from the two tables.

Figure 8.15: Comparing Columns in the Two Insurance Example Tables

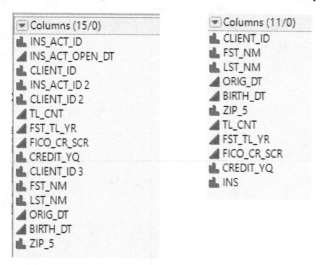

The task now is to combine the tables by adding rows from the non-insured table to the first table. This is precisely what the **Concatenate** command does.

1. Make the table of customers with insurance the active table. It doesn't actually matter which one we choose first, but this was the first table we created.

2. In the **Concatenate** dialog, click on the table of non-insured clients (**Sample of NON_INS_2028** in this example). Under **Data Tables to be Concatenated**, click **Add**.

    We have two options with respect to the combined data. We can append it to the first table, which is more efficient in terms of memory, or we can create a new table. Because neither of these tables is especially large, let's just make a new table.

3. Check **Create source column** to create a dummy variable for insured and non-insured clients, and provide a name for the new output table. We chose **All_Insurance**. (See Figure 8.16.) Then click **OK**.

Figure 8.16: Concatenating the Two Insurance Client Tables

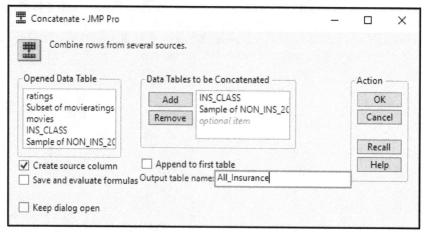

The result is a third table with 4,056 rows (the sum of the source tables) and 18 columns. We could then proceed to develop a classification model to identify likely prospects.

Query Builder for Tables

New in JMP 13 is the ability to use **Query Builder** for JMP tables, much as for ODBC and SAS data. Because we've worked with Query Builder in two prior chapters, the treatment in this chapter is brief and will highlight just a few additional features in two examples. First we'll revisit the MovieLens data from the early portion of this chapter, and then we'll combine data from two disparate examples of prior chapters: the World Development Indicators and Olympic Medals data sets.

Back to the Movies

The chapter began with a task of combining the ratings and movie data tables from MovieLens. Through the course of this chapter, we've made some minor modifications to column properties in the **ratings** and **movies** tables. But in this example, we'll go back to the original raw states of the data and re-open the originally saved tables. We'll create a four-column output table for films that were released in 1977, 1994, and 1999 just as before.

1. After opening both tables, modify the modeling type of **movieid** to **Ordinal**. Do the same for **userid** in **ratings**.
2. Make **movies** the active table.
3. Select **Tables ▶ JMP Query Builder**.
4. Choose **ratings** as the Secondary table and click **Build Query**.
5. In the next dialog, choose four columns to include: **movieid** (t1), **title**, **userid**, and **rating**.
6. Within the table of **Included Columns**, highlight **t1.title** and click the filter icon (looks like a funnel).
7. As shown in Figure 8.18, under **Filters**, click the red triangle next to **t1.title**, and select **Filter Type ▶ Contains**. Then type 1977 in the box next to **Contains**.

Figure 8.18: Filtering to Find Release Years

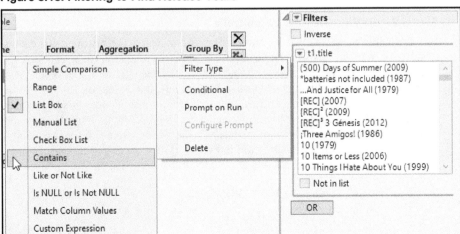

8. We now need to add similar conditions for 1994 and 1999. Click the **OR** button below this first condition. To see another way to add a filter, click on **t1.title** either in the **Available Columns** or the **Included Columns Variable Name** lists, and drag it to the new box marked **Add Columns or Filters**.

9. Again, click the red triangle next to the new **t1.title** filter and repeat the above steps for 1994 and 1999.

10. Run the query.

Here, in rather short order, we've joined and filtered the two tables within one JMP platform, rather than first joining and then subsetting or filtering. Moreover, Query Builder places the JSL and SQL code within table variables in the output table. As often happens, JMP provides several alternative paths to the same end and choices of the most efficient and appropriate method for a specific project.

Olympic Medals and Development Indicators

One benefit of the availability of reliable data from many sources is that it is easy to combine data in creative ways to investigate a question of interest. For example, articles by Bernard and Busse (2004) and Reiche (2016) analyze factors that contribute to Olympic success in some countries. One might start from these studies to develop testable hypotheses about why countries have varying degrees of success in the Olympic Games (as measured by medals won). In Chapter 5, we extracted a data table of Olympic medals data for summer games beginning in 1960, and in Chapters 4 and 7 we worked with the U.N. World Development (WDI) Indicators data.

We can use Query Builder again to see some of the possibilities and obstacles that can arise in a case like this. Let's suppose for simplicity that we want to extract some candidate explanatory variables from the WDI table, including the population, GDP, and a measure of health such as life expectancy. In addition, we'll include some columns that relate to gender parity in nations. WDI offers numerous series that could be useful, but this will suffice for present purposes.

Recall the structure of each table when we last worked with them. We had saved summer Olympics data since 1960, and we had made a wide form of the WDI data with 1,420 data series from 248 nations and country groups from 1960 through 2015. Let's try to combine those tables, extracting just the few columns and moderately few rows. Let's also sequence the rows of the new table first by year of the Games, and then by country.

More Wrangling before the Query

Before combining data tables, we should first recognize that the raw medalist data is not quite what we want for the project. The unit of analysis is the athlete in each separate event. But we eventually want to tally the number of medals won by each nation for each edition of the games. Moreover, there are individual events and team events; the winner of, say, the women's 100-meter freestyle swim earns one medal for her country. The successful women's soccer team also wins one medal, but 18 athletes carry home a medal – and appear in the imported data table. Hence, we want to aggregate the data to count medals per country per event. This was accomplished in a fairly routine way with **Tables ▶ Summary**. We'll work further with aggregating data in Chapter 11, and demonstrate the use of this valuable platform here.

1. Open the **Medalists since 1960** data table.

2. Select **Tables ▶ Summary**.

3. In the **Summary** dialog, first highlight the Medal column, and click **Statistics**. Choose **N**.

4. Select **NOC** and **Edition**, and click **Group**.

5. Make **Gender** the **Subgroup** in case we later want to analyze women's and men's medals separately.

6. Give the output table a name like **Medals by Country & Year**, uncheck the Link to original data table, and click **OK**.

Figure 8.19: Aggregating Data using Table ▸ Summary

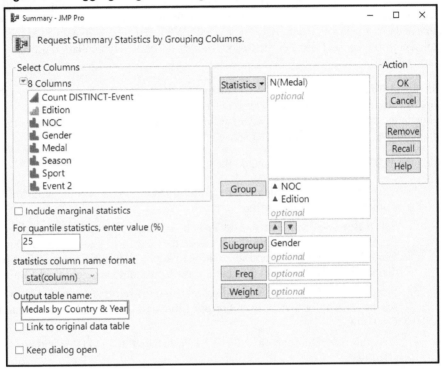

This will create a table with 5 columns and 736 rows. The first two columns bear the names from the original table, and JMP assigns the remaining three column names.

7. Let's rename the third one more usefully as **TotMedals**. Now we have a table showing the number of medals won by those countries winning medals in each of the summer games since 1960. We've also tallied men's and women's medal counts separately.

Additional Complications

Before combining the new **Medals by Country & Year** data with the WDI data, we should note three more wrinkles to iron out. First, there are IOC countries that don't win medals in any particular games. Second, there are nations in the WDI list that do not send teams to the Olympics. And third, the World Bank and the International Olympic Committee each have their own system of country abbreviations. So, for example, in the WDI data Bulgaria is coded as BGR. To the IOC, it is BUL and during the study period its athletes won medals.

Because we are interested in the drivers of success in the Olympics, we surely want to include participating countries with zero medals in any given year, but we should exclude countries that do not take part in the modern Olympic movement. Hence, we want to be sure to join our tables in such a way as to preserve all WDI countries with NOC codes and drop the rest.

Back in Chapter 6, we did gather a table of NOC codes from the International Olympic Committee. That table is called **List_of_IOC_country_codes 2**. Open it now. Now we need a table that includes both the World Bank and the IOC coding.

Fortunately, there are sources that list countries along with various standard coding schemes. One such source is Frictionless Data, which has just such a table on their website (frictionlessdata.io). Because we've already demonstrated importation from websites, we'll note that we read in this table and use it below to match IOC records with WDI data. The JMP data table for this book is called **country-codes.JMP**. For the curious reader, the entire reconciliation process is covered in Carver (2016).

Finally, as we prepare to combine WDI and medal-count data, it will be helpful to subset the WDI data by filtering out non-Olympic years. Later in the Query Builder we will select columns that relate to income, wealth, health, gender parity, and population size. There are several ways we might do this, but let's create another subset.

With the **WDI Wide** data table open, do the following:

1. Change the **Year** variable modeling type to Ordinal. This will facilitate the selection of summer Olympic game years.
2. Select **Rows ▶ Data Filter**. Add **Year** as a filter. As demonstrated earlier change the list of years to a check box list, and check the boxes every four years beginning with 1960: 1960, 1964, 1968, and so on through 2008. The WDI data continues past 2008, but that is where the Olympics data table ends. You should have checked 13 boxes.
3. Select **Tables ▶ Subset**. We will subset using the selected rows, and keep all the default settings except for assigning an **Output table name** like **WDI Wide Olympic Years**. Click **OK**.
4. Save this new table, and close the WDI Wide data filter and table without saving the row selections.

At this point, we want to work with four different data tables: **Medals by Country & Year**, **WDI Wide Olympic Years**, **country-codes**, and **List_of_IOC_country_codes 2**. If they are not already open, please open them now and close other tables. Select **List_of_IOC_country_codes 2** as the active table and proceed to build the query.

Selecting Tables and Key Columns for the Query

We will want to join the tables using the combination of the country code and the year, both of which have *different* names in the WDI and table of medals. Also, we want to be sure to include all Olympic nations, not just the medal-winners. We can accomplish this as follows.

1. Select **Tables ▶ JMP Query Builder**. Make the query table the primary source.
2. **Ctrl**-click to highlight the remaining three tables and add them all to the **Secondary** tables list.

   Note first that each table's name has a table number next to it, such as t1, t2, and so on. JMP is identifying each table with a number for use in building the query. Note also that each of the three secondary tables has a red x next to it. JMP can't make a best guess about columns to join because there are no matching column names. We need to specify the matches.

At this point, the Query Builder dialog looks like Figure 8.20.

Figure 8.20: Query Builder after Identifying Primary and Secondary Tables

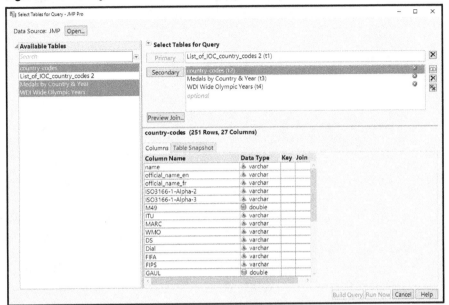

3. Let's first establish the matches between IOC codes and the identifiers that the World Bank uses in the WDI table. Highlight the **country_codes (t2)** item in the **Secondary** list; right-click; and choose **Edit Join**. This opens the **Add Condition** dialog.

4. In the **Left Column**, select **Code**, and in the **Right Column** scroll down until you can find and highlight **IOC** and click **Next**. (Figure 8.21)

Figure 8.21: Establishing Join Columns

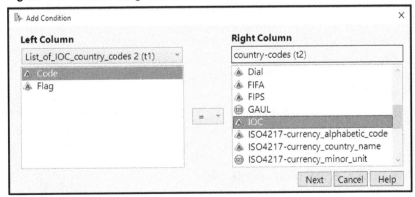

5. This opens another dialog (Figure 8.22). We want to map NOC codes to the country codes list to match with the WDI table. The default is a left outer join that would use all entries in the left column and omit those not found in the list of all country codes. All countries from the IOC list should be present in both tables.

6. So uncheck the box in the upper left of the dialog (**Include non-matching rows form List_of_IOC_country_codes 2**). Then click **OK**. In the list of secondary tables, the red x has been replaced by an inner join Venn diagram.

Figure 8.22: Choosing an Inner Join

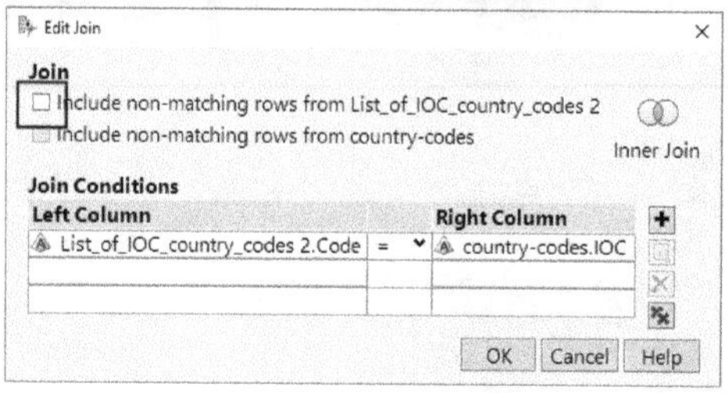

7. Highlight country-codes (t2), right-click, and choose Swap with Primary Table.

8. Now select WDI Wide Olympic Years (t4) in the secondary tables list, right-click, and choose Edit Join again.

9. In the next dialog (refer back to Figure 8.21), match ISO3166-1-Alpha-3, which is the standard country coding scheme used in the WDI data, to Country Code, and click Next.

10. Again, in the Edit Join dialog, we want an inner join. So uncheck the box in the upper left and click OK.

11. Again, select WDI Wide Olympic Years (t4), right-click, and make it the primary table.

12. Select Medals by Country & Year (t3), right-click, and choose Edit Join one last time. We now want to link the World Bank and Olympics data by both country codes and by years.

13. In the left column, choose Country Code; in the right column, choose NOC and click Next.

14. This time in the Edit Join dialog (refer to Figure 8.22), we do want a left outer join. That is, we want all Olympic countries, regardless of whether they won medals. This join is more complex than the prior ones, so do not move on yet.

15. Finally, we want to add one last condition, mapping years of summer games with WDI data. Next to the Right Column heading, there is a plus sign button. Click it to re-open the Add Condition dialog.

16. In the dialog, select Year on the left side, select Edition on the right, and then click OK.

17. In the Edit Join dialog, you will now see two join conditions. Click OK and then OK in the Select Tables for Query launch window.

Building the Query: Column Selection

Thus far we have identified the logical pathways that link the four tables together for this query, and have implicitly filtered out some rows via inner joins. Now it is time to select the columns to include as well as a sequence for the resulting query table.

Recall that the purpose of the investigation is to identify nation-specific factors that might account for variation in number of medals won at the summer Olympics. The earlier discussion hypothesized that variables related to population size, income, wealth, health, and gender equity might have predictive value. At this point, from the many available columns, we will specify a relatively small subset for those nations that do participate in the games in those years when Summer games have been held since 1960.

1. At the bottom of the **Select Tables for Query** dialog, click **Build Query**.

2. This opens another launch window, shown partly completed in Figure 8.23. In this window, we select the columns we want from the list of available columns in the lower right. In the top **Query Name** box, type `Olympics Query`.

Figure 8.23: Starting to Select Columns for the Olympics Query

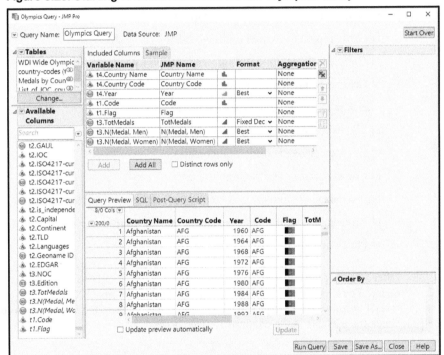

3. For this query, we want to select the following 26 columns, in this sequence. Select one or more column names and click **Add**. We can take advantage of the search box just below the **Available Columns** title. It will be helpful to widen the entire dialog as well as the list of available columns. Take your time and proceed carefully.

   ○ t4.Country Code

   ○ t4.Country Name

   ○ t4.Year

   ○ t1.Code

   ○ t1.Flag

   ○ t3.TotMedals

- ○ t3.N(Medal, Men)
- ○ t3.N(Medal, Women)
- ○ t4.GDP per capita (constant 2005 US$)
- ○ t4.GINI index (World Bank estimate)
- ○ t4.GNI per capita (constant 2005 US$)
- ○ t4.GNI per capita growth (annual %)
- ○ t4.Adjusted net national income per capita (current US$)
- ○ t4.GNI per capita, PPP (constant 2011 international $)
- ○ t4.GNI, Atlas method (current US$)
- ○ t4.GNI, PPP (constant 2011 international $)
- ○ t4.Gross capital formation (current US$)
- ○ t4.Gross enrolment ratio, primary, gender parity index (GPI)
- ○ t4.Gross enrolment ratio, secondary, gender parity index (GPI)
- ○ t4.Health expenditure per capita (current US$)
- ○ t4.Life expectancy at birth, female (years)
- ○ t4.Life expectancy at birth, male (years)
- ○ t4.Life expectancy at birth, total (years)
- ○ t4.Percentage of students in secondary education who are female (%)
- ○ t4.Percentage of students in secondary general education who are female (%)
- ○ t4.Population, total

4. After selecting the columns, click **Update** in the lower right to generate a preview of the query. You should see 26 columns. JMP will show the first 100 rows in a long query like this. If you are satisfied, click **Run Query.** If not, inspect the **Included Columns** to check the list against the columns shown above.

The Query Table

Part of the resulting query table is show in Figure 8.24. It has 26 columns and 2,483 rows corresponding to 13 years and 191 Olympic nations for which we have WDI data. We will work with this table in several of the coming chapters, so won't explore it very much here. Do note, though, that its full pedigree is recorded in the table variables pane of the data table window. We have the SQL script that generated it as well as all of the related prior scripts. This result is fully reproducible.

Figure 8.24: The Query Result

	Country Name	Country Code	Year	Code	Flag	TotMedals	N(Medal, Men)	N(Medal, Women)	GDP per capita (constant 2005 .
1	Afghanistan	AFG	1960	AFG			·	·	·
2	Afghanistan	AFG	1964	AFG			·	·	·
3	Afghanistan	AFG	1968	AFG			·	·	·
4	Afghanistan	AFG	1972	AFG			·	·	·
5	Afghanistan	AFG	1976	AFG			·	·	·
6	Afghanistan	AFG	1980	AFG			·	·	·
7	Afghanistan	AFG	1984	AFG			·	·	·
8	Afghanistan	AFG	1988	AFG			·	·	·
9	Afghanistan	AFG	1992	AFG			·	·	·
10	Afghanistan	AFG	1996	AFG			·	·	·
11	Afghanistan	AFG	2000	AFG			·	·	·
12	Afghanistan	AFG	2004	AFG			·	·	240.184?
13	Afghanistan	AFG	2008	AFG		1	1	0	294.2381
14	Albania	ALB	1960	ALB			·	·	·
15	Albania	ALB	1964	ALB			·	·	·
16	Albania	ALB	1968	ALB			·	·	·
17	Albania	ALB	1972	ALB			·	·	·
18	Albania	ALB	1976	ALB			·	·	·
19	Albania	ALB	1980	ALB			·	·	1792.818
20	Albania	ALB	1984	ALB			·	·	1792.655
21	Albania	ALB	1988	ALB			·	·	1742.371
22	Albania	ALB	1992	ALB			·	·	1094.275
23	Albania	ALB	1996	ALB			·	·	1645.577
24	Albania	ALB	2000	ALB			·	·	1985.900

As a preview of coming topics, it is immediately obvious that there is a sizeable issue of missing data here. Notably, in most years, relatively few participating countries earn any medals at all. Because they were missing from the medalist table, those observations come in here as missing rather than 0s. Moreover, many of the WDI observations are also missing for various reasons—though in those cases *missing* is decidedly different from *zero*. The challenges of missing data and other hurdles are the subjects of coming chapters.

Conclusion

This chapter has presented several ways to combine data from multiple JMP tables and to select rows and columns for later analysis. We've covered a lot of ground, but the variations on these basic techniques are extensive and go beyond a single chapter. We've seen a bit more of the versatility of Query Builder, and also encountered a few of the types of data preparation that form the subject of Part III of this book. In the next chapter, we'll look at some ways to explore and triage a data table looking for dirty data problems lurking beneath the surface.

References

Bernard, A, and M. Busse. 2004. "Who wins the Olympic games: Economic resources and medal totals." *The Review of Economics and Statistics,* February 86(1): 413–417.

Carver, R. 2016. "Speeding up the dirty work of analytics." *JMP Discovery Summit 2016*, Cary NC.

data.okfn.org. 2016. Comprehensive country codes: ISO3166, ITU, ISO4217 currency codes and many more. Available at http://data.okfn.org/data/core/country-codes.

GroupLens. 2016. "MovieLens Latest Datasets (Small)." Available at https://grouplens.org/datasets/movielens/latest/.

Hill, Eric. 2015. SAS Institute white paper. "Query Builder: The New JMP 12 Tool for Getting Your SQL Data into JMP." Available at https://www.jmp.com/content/dam/jmp/documents/en/white-papers/query-builder-jmp-12-107669_0515.pdf.

MovieLens. 2016. https://movielens.org.

Nashawaty, Chris. 2009. "Which was the best year for movies: 1977, 1994, or 1999?" Available at http://www.ew.com/article/2009/08/05/which-was-the-best-year-for-movies-1977-1994-or-1999.

JMP. 2015. *JMP Discovery Summit 2015*, San Diego CA.

Reiche, D. 2016. "Want more Olympic medals? Here's what nations need to do to win." *Washington Post,* Aug. 3. Available at https://www.washingtonpost.com/news/monkey-cage/wp/2016/08/03/want-more-olympic-medals-heres-what-nations-need-to-do-to-win-in-rio/

SAS Institute Inc. 2016. *Using JMP*. Cary NC: SAS Institute Inc.

Chapter 9: Data Exploration: Visual and Automated Tools to Detect Problems

Introduction

In Part II of this book, we've discussed many of the issues that arise on the path from disparate raw data sources to a consolidated JMP data table. Earlier chapters have discussed issues such as adjusting data types or modeling types "on the way into" JMP. Once the data have all been assembled and transferred into JMP, there are typically further obstacles to navigate. Some of them might be obvious at the outset, and others need to be discovered through a combination of familiarity with the substance of the particular project and familiarity with the statistical methods to be used. This chapter deals with the triage phase of preparing data for modeling.

This chapter explains several types of issues to expect, and illustrates some of the ways that JMP helps reveal problem areas. Rather than explore the raw data without a purpose or agenda, analysts should look for particular pitfalls or rough edges to smooth out. This chapter suggests where and how to look. It also points ahead to the next few chapters that demonstrate common approaches to resolving the problems.

Common Issues to Anticipate

It is important to recognize that the details of data preparation will vary depending on the analysis to be performed and the data types involved. Nevertheless, when you know which rocks to turn over, the process goes more efficiently. Kandel *et al.* (2011) describe this phase of a project as an iterative process that might send the investigator back for changes in the extraction stages previously discussed, and might surface the need for more or different data. The goal is to make data *usable, credible,* and, *useful. Usable* refers to getting them into a form that is suitable for the intended analysis tools. We've taken the first step toward usable data by building and saving JMP data tables, but the intended analysis might require additional steps.

Credible data are representative of the target process for the purposes of the study. This is not necessarily the same as the traditional statistical meaning of representativeness, but connotes accuracy and suitability for the particular inquiry.

Finally, *useful* data are necessarily usable and credible and must also be "responsive to your inquiry" (Kandel 2011, p. 272). In the framework suggested by these authors, the result of a successful data wrangling effort also includes an "editable and auditable transcript of transformations coupled with a nuanced understanding of data organization and data quality issues."

At one level, these criteria direct attention to questions about whether the data table contains the "right," or at least acceptable, columns and rows. Do the variables (features, factors, and columns) adequately capture the model constructs to serve the current purpose? Do the observations (cases, subjects, instances, and rows) suitably represent the subjects being investigated or modeled? If a study is about organisms, do the rows represent organisms? If the study is about surgical procedures, do the rows contain measurements obtained during operations? These substantive questions speak to the analyst's domain knowledge and understanding of the nature of the inquiry.

Some of these matters might appear obvious, but when a study relies on a compilation of data originally gathered for various reasons, and then assembled and repurposed for a given study, it is important to ask these questions about usability, credibility, and usefulness. Beyond the problem areas just cited are other forms of "dirtiness" to consider. Kim *et al.* (2003) have offered a comprehensive taxonomy of dirty data. Drawing on Kim and other sources, this chapter focuses on these specific obstacles:

- *Missing Data*: After assembling a data table, you find empty cells. How prevalent are they? Do they occur randomly or with meaningful patterns? If a model requires complete cases for all variables in the model, we can lose entire cases if there are any gaps at all across the variables. Might we infer meaning from a blank cell? Might we sometimes be able to impute (estimate) a suitable replacement value for a blank and thereby retain the observation?

  Some data sources use an arbitrary missing data code like -99; a data dictionary is needed to identify such practices. Others might use a string like "NA", which JMP would recognize as character rather than numeric data, so that the impact of missing data might also include the need to correct the data and modeling type for a column.

- *Outliers:* Which columns contain extreme observations that might be erroneous or might signal an observation that does not come from the target population? Are there outlying values that are accurate and hence should be retained, but that can potentially distort a visualization or the results of the planned data analysis?

- *Data errors:* Are there individual values that are simply incorrect? De Veaux and Hand (2005) cite various possible causes of data errors including miscalibration of instruments, human transposition of digits, mislocation of decimal points, and changes in measurement protocols and procedures. Some such errors might be difficult or impossible to detect at scale, but the initial investigation should seek them out.

- *Skewness:* Some, but not all, methods rely on data that are approximately normal in shape, or at least not severely skewed. Is that the case in the current data table?

- *Non-linear relationships*: Some, but not all methods fit models that are linear. Are there columns in the table that appear to be related but in non-linear fashion?

- *Too-good-to-be-true relationships*: Particularly with foreign (that is, non-internal) data sources, some columns might be computed from others and hence have remarkably strong associations with the originating columns. But those relationships are non-informative. Alternatively, columns might be highly correlated because they represent slightly different approaches to the measurement of a single construct, and hence would be redundant in a model.

On the Hunt for Dirty Data

JMP users are probably familiar with several approaches to data exploration to find errors and unusual observations. In this section, we demonstrate five sensible ways to start. These examples are intended to be illustrative as opposed to comprehensive, because the quirks of any individual data table are so varied. This section illustrates five JMP features that are especially relevant at this stage.

- **Analyze ▶ Distribution** for a quick view of distributions, shapes, outliers, and N missing. Use this to spot skewness, data errors, incorrect data types, and outliers.

- **Cols ▶ Column Viewer** for numerical summaries. Non-graphical, but clearly identifies missing cases. You can invoke the **Distribution** report from here.

- **Analyze ▶ Multivariate Methods ▶ Multivariate** computes correlations and creates a scatterplot matrix for continuous variables. Also useful for identifying outliers, non-linear relationships, and near-perfect (or obvious) relationships, and helpful with feature selection depending on the modeling strategy.

- **Analyze ▶ Screening ▶ Explore Outliers** provides four different methods for identifying outliers in the data.

- **Analyze ▶ Screening ▶ Explore Missing** provides numerical and visual descriptions of missing values as well as providing two imputation methods, to be discussed further in Chapter 10.

In the next five sections, we look more closely at each of these methods. All of the examples continue with the combined Olympic medals data (**Olympics Query.jmp**) from the previous chapter. Recall that the table has 26 columns and 2,483 rows. In most of the examples, we'll examine the first several continuous columns. In a full study, we'd want to investigate all columns individually or in meaningful groups, but the goal here is to demonstrate the approach and methodology.

Distribution

For a quick overview of univariate distributions, select **Analyze ▶ Distribution** and choose all columns (for tables with many columns, you might wish to select several at a time). For this example, we'll work with the response variable (number of medals), and four measures related to national income. Figure 9.1 shows the results for these five columns, which easily fit on a single screen. It is clear in this display that three of the variables are very right-skewed. Though hard to see in this image, except for the second column (GDP per capita; see inset), approximately half or more of the observations are missing. There are outliers visible in the box plots for four of the columns.

Figure 9.1: Distributions of Continuous Columns

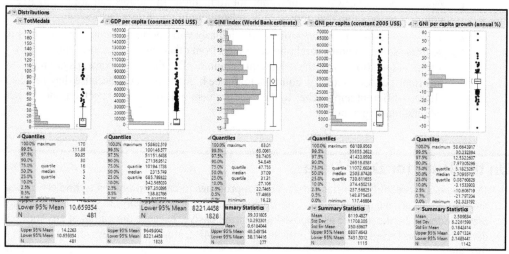

In this particular set of data, the only categorical variables are the country identifiers and names, so there is little point in displaying their distributions. In other contexts, you should check for inconsistent coding and for categories that might well be grouped together as "other."

Throughout the data table, many of the variables are strongly skewed. For example, most nations have relatively low per capita wealth, with a very long right tail to the distributions. Before proceeding with imputation or modeling, we might well want to apply a log or other transformation. Because some variables are percentages and others are dollar amounts, we might want to standardize columns. These are typical of the data preparation process, and JMP makes them easy to accomplish either within the data table or as a local transformation within an analysis platform.

Columns Viewer

Since JMP 11, users have been able to use **Cols ▶ Columns Viewer** to obtain simple summaries and then to examine distributions. Using the same columns as in the prior example, we would invoke the column viewer, highlight the relevant columns, and click **Show Summary**. For continuous data, the result will be similar to that shown in Figure 9.2.

Figure 9.2: Columns Viewer Summary

Summary Statistics

5 Columns [Clear Select] [Distribution]

Columns	N	N Missing	Min	Max	Mean	Std Dev
TotMedals	481	2002	1	170	12.4	19.906483332
GDP per capita (constant 2005 US$)	1828	655	73.828772502	158602.51925	8935.6750233	15570.042846
GINI index (World Bank estimate)	277	2206	16.23	63.01	39.331805054	10.292300573
GNI per capita (constant 2005 US$)	1115	1368	117.46884012	68189.956342	8119.4827138	11708.305067
GNI per capita growth (annual %)	1142	1341	-52.32319213	58.664391672	2.5098340488	6.2261597951

Here we see explicit counts of missing values, as well as minimum and maximum values along with means and standard deviations. These summaries can help with checking for reasonable values and to signal which columns need attention due to missing observations. The platform also includes a button to open the **Distribution** report identical to the one in Figure 9.1.

For this example, we also selected the three nominal variables **Country Name**, **Country Code**, and **Code** (NOC), and clicked **Show Summary**. The report shows three variables that all have 191 categories (Figure 9.3), but the Distribution report (Figure 9.4) clearly reveals the mismatched coding issue that we resolved in the prior chapter.

Figure 9.3: Columns Viewer Summary for Categorical Columns

Summary Statistics

3 Columns [Clear Select] [Distribution]

Columns	N Categories
Country Name	191
Country Code	191
Code	191

Figure 9.4: Distribution of Categorical Columns (Detail)

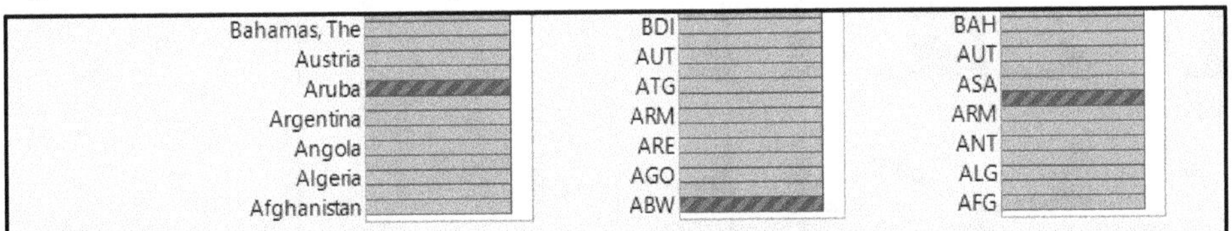

Frequencies

Level	Count	Prob
Afghanistan	13	0.00524
Albania	13	0.00524
Algeria	13	0.00524
American Samoa	13	0.00524
Angola	13	0.00524
Antigua and Barbuda	13	0.00524
Argentina	13	0.00524
Armenia	13	0.00524
Aruba	13	0.00524
Australia	13	0.00524
Austria	13	0.00524

Frequencies

Level	Count	Prob
ABW	13	0.00524
AFG	13	0.00524
AGO	13	0.00524
ALB	13	0.00524
ARE	13	0.00524
ARG	13	0.00524
ARM	13	0.00524
ASM	13	0.00524
ATG	13	0.00524
AUS	13	0.00524
AUT	13	0.00524

Frequencies

Level	Count	Prob
AFG	13	0.00524
ALB	13	0.00524
ALG	13	0.00524
ANG	13	0.00524
ANT	13	0.00524
ARG	13	0.00524
ARM	13	0.00524
ARU	13	0.00524
ASA	13	0.00524
AUS	13	0.00524
AUT	13	0.00524

There is nothing to repair or clean here; we reconciled the codes during extraction. However, this does illustrate another potential issue: If we later want to graph selected countries, in, for example, alphabetical order, the order will differ depending on whether we use the WDI country code, NOC codes, or country names.

Multivariate (Correlations and Scatterplot Matrix)

Much as looking at univariate distributions is a natural first step in diagnosing data cleaning issues, it is also wise to explore bivariate patterns as well—depending on the nature of the inquiry. In this Olympics example, we have a dependent/response/target variable that we intend to model using several constructs identified by prior research in the field. Hence, it would be sensible first to see how the measures for each construct covary with the number of medals won and with each other.

We might need to repeat this exercise later in the process after addressing issues of missing observations and other data irregularities, but an initial triage review should proceed as follows.

1. Select **Analyze ▶ Multivariate Methods ▶ Multivariate**.
2. Select a set of variables to inspect. Initially, we've choose the total number of medals won as well as the first four WDIs related to national income and wealth. For the sake of creating a clear image here, in the lower left of the dialog, we've chosen **Upper Triangular** as the **Matrix Format**.

The resulting report consists of two parts. At the top is a correlation matrix, and below it is a scatterplot matrix (Figure 9.5). The top row of the scatterplots shows the relatively weak bivariate relationships between the number of medals won and the wealth measures over all years. In four of the five graphs in that row, there is one clear outlier (the United States)/ In the second graph, there is a cluster of several outlying countries.

Figure 9.5: Scatterplot Matrix

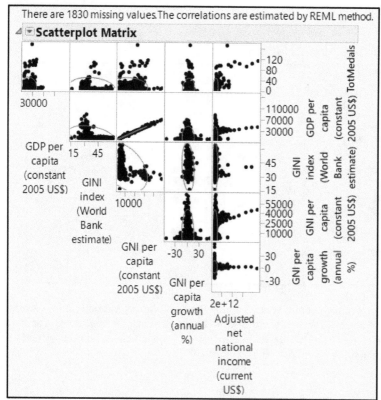

The fourth graph in the row might lead us to think of dropping **GNI per capita growth (annual %)** from further consideration, as having low predictive value. With its large number of missing values (seen earlier in the univariate reports), it won't add much to our investigations.

In addition, we see that some candidate variables are redundant. Not surprisingly, gross domestic product (GDP) per capita and gross national income (GNI) per capita in constant USD have a near-perfect linear relationship ($r = 0.9988$). They fundamentally measure the same thing. Earlier we found that there are only 655 missing values for GDP, but more than twice that number of missing observations for GNI. We could drop GNI from further modeling or apply a technique like Principal Components analysis to combine the two variables into one.

More Tools within the Multivariate Platform

In addition to the visual inspection, the **Multivariate** platform offers a large number of further diagnostic tools and interventions for addressing data imperfections. By clicking the red triangle next to **Multivariate**, we see an extensive menu of capabilities. The two topmost blocks provide numerical measures and visualizations of associations. The next two groups include commands for further analysis of commonalities and differences across groups of columns as well as one approach to missing value imputation. This section provides an overview of the three commands highlighted in Figure 9.6.

Figure 9.6: Multivariate Platform Menu

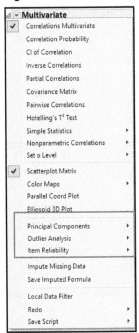

Principal Components

This data table illustrates one common situation: We have several measures of a conceptual construct, economic well-being. No single variable alone captures a large amount of the variation in the target variable. Ultimately, we will prefer a model that is parsimonious with the number of independent variables used. Principal Components Analysis (PCA) is one approach to dimension reduction—that is, reducing the number of variables for model-building. PCA creates a relatively small number of separate linear combinations of several numeric variables in a way that captures as much of the variation in the original variables as possible.

At the exploratory stage, PCA is a way to ask "how many distinct underlying dimensions are being measured in this set of columns?" In this example, if our five variables truly all measure the same thing, then PCA should reveal a single principal component. If the analysis were to find more components, we'd want to think about the more subtle differences in measures.

A full treatment of PCA is beyond the scope of this chapter, but we'll take it up later in Chapter 11. Interested readers should skip ahead, or consult the relevant section in the *Multivariate Methods* book in the **Help** menu.

Outlier Analysis

Scatterplots show bivariate outliers in an intuitive way, but we might also want to identify and investigate cases that deviate from a *multivariate* framework. Outlier Analysis presents several alternative measures to locate a row that deviates dramatically from the multivariate pattern, relative to other rows.

For example, consider the variables illustrated above. The Gini index or coefficient is a measure of income inequality in a country. A coefficient of 0 indicates perfect equality and 1 perfect inequality. A univariate exploration might identify a country with an extraordinarily low Gini index. In a bivariate analysis, we might look at the relationship between (for example) Gini index and per capita GDP and look for countries with exceptional Gini values, given their level of GDP per capita.

The outlier analysis extends the very same logic to more variables, looking for rows (countries) that depart from the correlation structure of the selected columns. The platform provides a choice among three measures: Mahalanobis distance, jackknife distance, and T^2 statistic. For readers new to these measures, consult the Multivariate Platform Options in the JMP *Multivariate Methods* book under the **Help** menu.

Item Reliability

Reliability refers to the consistency of a set of measures with respect to a single construct. To use a different Olympics-related example, in an event with multiple judges scoring an athlete's performance (think diving or figure skating), we would want to know how consonant the various judges were with one another. This is the central issue of reliability, and the most common measure is Cronbach's α. As a general rule, values approaching 1 indicate a high level of confidence that the variables are providing a reliable measure.

Explore Outliers

Beyond visual inspection in the **Distribution** and **Multivariate** reports, the now-relocated **Explore Outliers** feature can be very helpful in devising a strategy suitable for the project at hand. It is now in the **Analyze** menu among the **Screening** platforms. Let's continue examining the wealth-related columns in this table, and expand our investigation to the response variables and all nine wealth columns (**GDP per capita (constant 2005 US$)** through **Gross Capital Formation (current US$)**).

1. Select **Analyze ▶ Screening ▶ Explore Outliers**.

2. Cast the columns of interest as Y, and click **OK**. This opens the dialog shown in Figure 9.7. Since there are multiple working definitions of "outlier," JMP offers four basic approaches to identifying extreme and unusual values. These are fully documented in the book *Fitting Linear Models*, found on the Help menu in JMP. We briefly summarize the distinctions among the four approaches here, and then illustrate use of the options.

Note that each approach offers several user-controlled options. Here the goal is to describe the four methods, drawing analogies to familiar contexts. The first two options use univariate measures on selected columns, and the second two are intended for multiple columns.

Figure 9.7: Explore Outliers Dialog

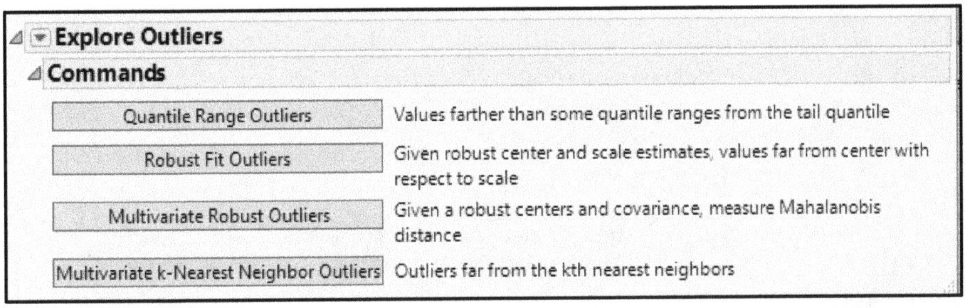

⊿ ▽ **Explore Outliers**	
⊿ **Commands**	
Quantile Range Outliers	Values farther than some quantile ranges from the tail quantile
Robust Fit Outliers	Given robust center and scale estimates, values far from center with respect to scale
Multivariate Robust Outliers	Given a robust centers and covariance, measure Mahalanobis distance
Multivariate k-Nearest Neighbor Outliers	Outliers far from the kth nearest neighbors

- **Quantile Range Outliers**: This method defines outliers as being points found beyond a multiple of an interquantile range specified by the user. The default values are 0.1 and 3, meaning that points beyond 3 times the difference between the 90th and 10th percentiles. This is analogous to use of a multiple of the interquartile range in the outlier box plot.

- **Robust Fit Outliers**: This method first computes robust measures of center and dispersion for the variable. It then defines outliers as those lying more than K times the robust dispersion measure from the robust center. This is analogous to classifying observations more than, for example, 3 standard deviations from the mean of a symmetric bell-shaped distribution.

- **Multivariate Robust Outliers**: This method and the next method refer to outliers from multi-column relationships for the columns selected. It calculates the Mahalanobis distances between each point to the robust center of a multivariate normal distribution, relative to the estimated correlations. In a bivariate context, this is akin to looking for points lying "far" from a linear pattern.

- **Multivariate k-Nearest Neighbor Outliers:** Finally, this method takes a fundamentally different approach. It computes the Euclidean distance of each individual point from its Kth nearest neighbor, where K is supplied by the user. Hence, rather than expressing distance from a center distance is relative to the nearest of a subset of points.

Of course, once outlying values have been identified there is the question of what should be done, and those judgments require domain knowledge. Depending on the context and the process by which the data were generated and captured, the most appropriate approach will vary. In some instances, outliers are caused by measurement error, instrument failure, or human error in recording or transcribing the data. In others, outliers are best handled by a variable transformation such as taking a logarithm. Chapter 11 provides a series of approaches, and here we'll look at the reports for the four options using the columns related to the total number of medals won and the economic well-being of the nations.

Quantile Range Outliers

Clicking on **Quantile Range Outliers** produces the report shown in Figure 9.8. Please note that the figure has been truncated on the right side. Also note that the **Commands** have been hidden, but will re-appear upon clicking the gray disclosure arrow.

Figure 9.8: Quantile Range Outliers Report

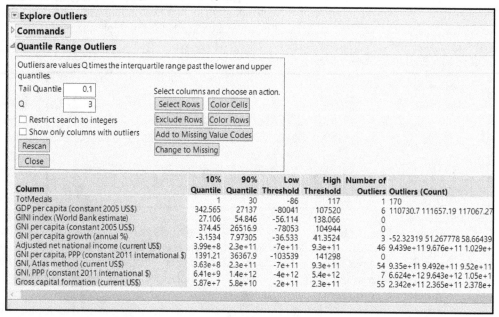

The approach here is described above. Let's examine the first selected column to see how this works. For **NMedals**, the interquantile range is $30 - 1$ medal $= 29$ medals. Multiply by 3 (the default value of Q) and get 87. The Low Threshold is $1 - 87 = -86$; similarly, the High Threshold is $30 + 87 = 117$. Only one country won more than 117 medals over the years, and the outlying value is 170. The right-most column of the report identifies the outlying values for each column.

The buttons in the upper panel of the report are largely self-explanatory, allowing for ways to highlight, select, or exclude the cells and rows. You can also treat the outlying values as missing if that is appropriate in the context of the investigation. After excluding or converting extreme values to missing, the thresholds will change. Click **Rescan** to re-evaluate.

You should exercise caution and judgment before too quickly discarding outliers or treating them as if they were missing. It is essential to understand why extreme values are extreme, and simple non-conformance with a larger pattern is no reason to delete cells. Chapter 11 has considerably more to say on this point. It is far better to temporarily exclude a row than to re-code it as missing.

Robust Fit Outliers

Continuing with the same example, we can open the hidden list of **Commands** and choose the second option of **Robust Fit Outliers**. In this illustration, I have selected all four available methods so that we can see the effects of different methods. As we see in Figure 9.9, robust fits tend to classify many more values as extreme in comparison to the interquantile range approach.

Figure 9.9: Robust Fit Outliers

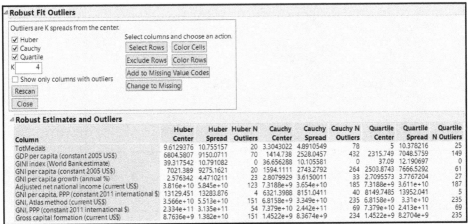

Column	Huber Center	Huber Spread	Huber N Outliers	Cauchy Center	Cauchy Spread	Cauchy N Outliers	Quartile Center	Quartile Spread	Quartile N Outliers
TotMedals	9.6129376	10.755157	20	3.3043022	4.8910549	78	5	10.378216	25
GDP per capita (constant 2005 US$)	6804.5807	9150.0711	70	1414.738	2528.0457	432	2315.749	7048.5759	149
GINI index (World Bank estimate)	39.317542	10.791082	0	36.656288	10.105581	0	37.09	12.190697	0
GNI per capita (constant 2005 US$)	7021.389	9275.1621	20	1594.1111	2743.2792	264	2503.8743	7666.5292	61
GNI per capita growth (annual %)	2.576342	4.4710211	23	2.8079929	3.6150011	33	2.7095573	3.7767204	27
Adjusted net national income (current US$)	3.816e+10	5.845e+10	123	7.3188e+9	3.654e+10	185	7.3188e+9	3.611e+10	187
GNI per capita, PPP (constant 2011 international $)	13129.451	13283.876	4	6321.3988	8151.0411	40	8149.7485	13952.041	5
GNI, Atlas method (current US$)	3.566e+11	5.513e+10	151	6.8158e+9	3.349e+10	235	6.8158e+9	3.31e+10	235
GNI, PPP (constant 2011 international $)	2.334e+11	3.135e+11	54	7.379e+10	2.442e+11	69	7.379e+10	2.413e+11	69
Gross capital formation (current US$)	8.7636e+9	1.382e+10	151	1.4522e+9	8.3674e+9	234	1.4522e+9	8.2704e+9	235

As noted earlier, one traditional approach would be to compute the mean and standard deviation of a column and then look for values more than two or three standard deviations from the center. This is sensible when variables are normally distributed. Cauchy and Huber are two robust methods that generate different measures of center and spread. The Quartile range approach considers the median to be the center, and the spread is the IQR divided by 1.35, which approximates one standard deviation in a normal distribution. Finally, K is the multiplier for the spreads.

Notice that the classification of extreme values is sensitive to the choices of K and the methods. The analyst needs to shoulder the burden of deciding on the most appropriate definition of outlier in the univariate sense.

Multivariate Robust Outliers

Just as we saw earlier in the section discussing the menu options on the Multivariate report, this option looks for rows that depart substantially from the correlational structure of the entire set of selected columns. There are three clearly exceptional values: one for China (labeled in Figure 9.10) and two for the United States. That said, though, there are quite a large number of observations lying above the upper threshold of distances.

Figure 9.10: Multivariate Robust Outlier Report of Mahalanobis Distances

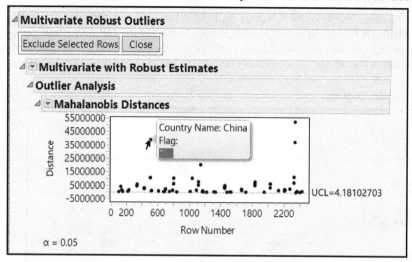

Unfortunately, this option does not include a button to select unusual rows. However, as with all JMP plots, we can use the Arrow tool to select points of interest within the graph and thereby select the rows in the data table. In this graph, 94 rows have Mahalanobis distances greater than the upper control limit of 4.18, and selecting any or all will identify them.

Multivariate *k*-Nearest Neighbors Outliers

Finally, choosing the final option first opens a small dialog asking you to specify a value of K. The default is 8, and that is the value used here. The report appears in Figure 9.11.

Figure 9.11: K Nearest Neighbors Report

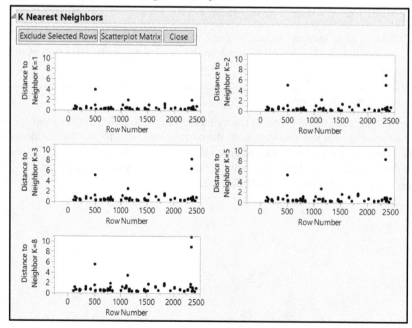

Notice that the pattern of points remains stable in this example as K goes from 3 to 5 to 8. Seemingly, the same rows are unusual, judging by their distance from a small number of nearest neighbors. As with all the methods demonstrated here, you might exclude the unusual rows. Also, as always, brushing across points selects the rows.

Explore Missing

Just below **Explore Outliers** on the **Screening** platform is the **Explore Missing** platform. To begin illustrating this feature, let's continue to look at the nine WDIs related to the wealth of the nations. Figure 9.12 shows the initial report on missing values for that group of columns. The information here is the same as the missing values shown earlier in the **Columns Viewer** output, though here we have additional columns.

1. Select **Analyze ▶ Screening ▶ Explore Missing**. In the dialog, select the nine wealth-related columns. This produces the report shown in Figure 9.12.

Figure 9.12 Missing Value Report on Wealth Measures

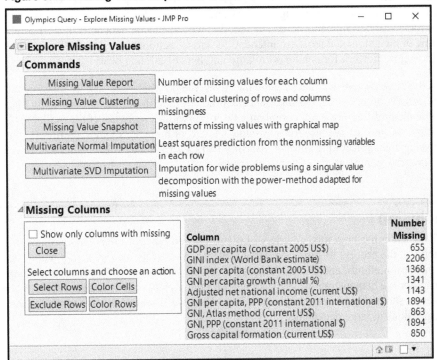

2. If we then click **Missing Value Clustering**, we get a hierarchical graph of missing observations by column and row. Because it is a tall narrow image, we've rotated the view in Figure 9.13. Blue areas show where observations are present, and red indicates missing observations. The idea in this diagram is to find groups of columns where missingness is least problematic and to know which rows are most impacted by missing data. While useful in its own right, it is particularly valuable if we decide to use one of the automated imputation methods that actually use different algorithms to replace blank cells with values. We'll come back to that in Chapter 10.

Figure 9.13: Missing Value Clustering

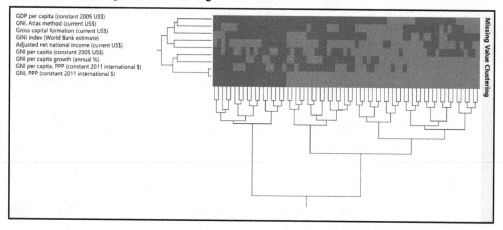

We've seen that we have many empty cells in our combined table, and JMP offers several tools that can help us deal with them efficiently. These tools are the exclusive concern of Chapter 10. Because the reasons for missingness vary and the demands of different projects differ, not all of the tools will be applicable here. The fact that we have panel data adds to the complexity since some of the automated tools are intended for cross-sectional samples. Still, the tools provide an efficient framework for coping with large numbers of missing observations.

As the dimensionality of a data table increases, the penalties imposed by missing cells grow. In the context of modeling, missing data can pinch us in multiple ways. For multivariable models, some methods require a full complement of observations rowwise. If the model for Y includes factors X1 and X2, we lose rows missing X1, X2, or both. This fact magnifies the impact of missing data. Selecting **Tables ▶ Missing Data Pattern** identifies the number of rows for which all possible combinations of columns are missing.

Conclusion

This chapter has presented a variety of platforms and functions that can automate or otherwise facilitate the process of surfacing data problems. Some rely on visualization and others on numerical methods, and all supplement the critical thinking of the analyst. In the next two chapters, we demonstrate ways of coping with the major problem areas of missing data, highly skewed samples, and extreme observations. Also, we'll see some ways to reduce the number of dimensions in large data tables.

References

De Veaux, Richard D. and David J. Hand 2005. "How to lie with bad data." *Statistical Science,* 20(3), pp. 231-238.

Kandel, Sean, Jeffrey Heer, Catherine Plaisant, Jessie Kennedy, Frank van Ham, Nathalie Paolo. Henry Riche, Chris Weaver, Bongshin Lee, Dominique Brodbeck, and Paolo. 2011. "Research directions in data wrangling: Visualizations and transformations for usable and credible data". *Information Visualization.* Vol 10(4), pp. 271-288.

Kim, Won, Byoung-Ju Choi, Eui-Kyeong Hong, Soo-Kyung Kim, and Doheon Lee. 2003. "A taxonomy of dirty data." *Data Mining and Knowledge Discovery,* 7(1), pp. 81-99.

McCormack, Don. 2015. "It's a dirty job, but someone has to do it." Presentation at *JMP Discovery Summit 2015*. Downloaded from https://community.jmp.com/t5/Discovery-Summit-2015/It-s-a-Dirty-Job-but-Someone-Has-to-Do-It/ta-p/23806.

SAS Institute Inc. 2016. JMP 13 *Fitting Linear Models*, Cary NC: SAS Institute Inc.

SAS Institute Inc. 2016. JMP 13 *Multivariate Methods*, Cary NC: SAS Institute Inc.

Shmueli, Galit, Peter C. Bruce, Mia L Stephens, and Nitin R. Pate. 2016. *Data Mining for Business Analytics: Concepts, Techniques, and Applications with JMP Pro*. Hoboken, NJ: John Wiley, Chapter 4.

Chapter 10: Missing Data Strategies

Introduction

Chapter 9 illustrated several ways to detect missing observations within a data table and pointed ahead at some strategies for dealing with this common issue. This chapter illustrates four basic approaches to the problem. Unfortunately, there is no single strategy that dominates all others. The best approach depends on the goals of the project, the data types involved, and the domain knowledge of the analyst. As with many other topics in this book, the examples presented here are intended to illustrate what you might do, and how JMP can help smooth the way forward.

The references presented at the end of the chapter will direct readers to other sources for further background. As in Chapter 9, the analysis of Olympic medal-winning runs through the chapter, but other examples are introduced to clarify and demonstrate basic concepts.

There are a few important ideas to consider at the outset of this discussion.

- Some statistical methods require complete cases (rows) of data to produce estimates. For example, linear regression models need values of the response variable and all factors for every row in a data table. Suppose we want to build a model to estimate Y from three X variables. If X_3 is missing for one subject and X_2 is missing for another, then both subjects are dropped from the analysis, which is likely to bias the results except under certain restrictive assumptions.

- In a given study, where data are being combined from several sources, missing data might be more or less troublesome for the ultimate user of the data than for the original constructor of the database (Rubin 1996). Depending on the study goals, the analyst might be able to side-step some missing values, or use an analytic technique that does not require complete cases. She might be able to impute values to fill in blank cells, or use a technique that draws meaning from the fact that a cell is missing.

- Unless there is reason to think otherwise, missing data is potentially problematic for two serious reasons: Missing observations can lead to biased estimates, and to misestimation of sampling variability. In plainer language, models built on a foundation of sparse or incomplete data sets can (a) systematically over- or under-estimate model parameters and (b) inaccurately account for uncertainty. Hence, statistical inferences will be unreliable and decisions based on such inferences can produce undesirable and unpredictable consequences.

- Although this chapter introduces several approaches to working with data tables that suffer from sparsity, there's no alchemy here. A data set with missing data is apt to be inferior to a complete record. These techniques compensate for the deficiencies in various ways, but do not render a gap-filled data table into one that is identical to one that was complete from the outset.

- Lastly, this book focuses on managing data. Therefore, in this chapter we'll consider methods of preparing the data table itself prior to analysis. We'll mention, but only sometimes illustrate ways in which some **analysis platforms** compensate for missingness while carrying out their work.

Much Ado about Nothing?

This section heading, borrowed from Horton and Kleinman (2007), is another way of asking, Why does the topic of missing data deserve an entire chapter? In the most general sense, it simply refers to blank or unrecorded cells within a data table. Sometimes we know or can surmise the reasons that led to empty cells, and sometimes we cannot. Sometimes, the very fact that an individual case has no value for a given column is informative in its own right.

As an introductory example, consider the "Youth Risk Behavior Surveillance System" that generated the data found in the JMP Sample Data table titled **Health Risk Survey**. This particular iteration of the survey was taken in 2005. Researchers at the U.S. Centers for Disease Control interviewed nearly 14,000 9th to 12th grade students asking a large number of questions about health and risky teen behaviors. The most current *User's Guide* (CDC 2016) for this sample data describes its methodology as follows:

The National Youth Risk Behavior Survey (YRBS) uses a three-stage cluster sample design to produce a representative sample of 9th through 12th grade students. The target population consisted of all public, Catholic, and other private school students in grades 9 through 12. A weighting factor was applied to each student record to adjust for nonresponse and the oversampling of black and Hispanic students in the sample.

This description points to two types of missingness from the outset: Some target populations are over- or under-represented due to the sampling method, due to the fact that some people decline to be interviewed, or perhaps due to the fact that they have stopped attending school. Therefore, there are individuals absent from the sample whom researchers would have wanted to include. Also, sampled individuals might have chosen to skip some questions while answering others.

These types of challenges can occur in any type of data collection. Huge data sets with many columns are infamous for their sparsity—an enormous number of columns with mostly blank cells. Why is sparsity a challenge? There are a number of issues of computations inefficiency, but for the purposes of this discussion, there are two big statistical problems to tackle: bias and unknown variability. These problems affect both the estimator and standard errors—meaning that inference is jeopardized.

The severity of the problems depends both on the extent of missingness relative to the sample size and the mechanisms giving rise to missingness. Similarly, the choice of practical responses to missing-data issues also depends on how and why those pesky blank cells occur in the first place. In cases where the reasons for missingness are well understood, it's no small matter to select the most appropriate approach to data preparation and analysis. When the underlying mechanism is not well understood, it is ultimately up to the analyst and subject-matter experts to decide on the assumptions that they are willing to make. There is no purely statistical rationale that is valid in all studies. This chapter lays out several methods that are appropriate depending on the type of missing data.

Little and Rubin (2002) differentiate among three types of missing data (also Gelman and Hill 2006; Penn 2007; and Sterne *et al.* 2009), which can be understood in the context of the YRBS example.

- **MCAR: Missing Completely at Random**. Blank cells occur essentially without any pattern, and the likelihood of missingness is the same for every case. Observed values provide no information about the missing values. Missingness does not depend on any observed data. A data value is missing because of a transcription or data entry error that might occur for any subject in the study. When it is justifiable to think that data are MCAR, complete case analysis is likely to be unbiased.
- **MAR: Missing at Random.** The probability that an observation of Y is missing is a function of observed variables that are completely reported. When data are MAR, analytic techniques that require complete cases might yield biased estimates. For example, if it is plausible to assume that students' propensity to skip a question about drinking behavior does not depend on other factors such as age, we might consider NAs for the drinking question to be MAR.
- **MNAR: Missing Not at Random**. If the pattern of missingness (not the values themselves, but the fact that they are missing) depends on unobserved data, then empty cells are missing not at random. In other words, "[e]ven after the observed data are taken into account, systematic differences remain between the missing values and the observed values." (Sterne, *et al.* 2009, p. 157). If, for example, the fact that a student engages in a non-reported illegal activity increases the probability that the student will skip a question about something else, the NAs would be missing not at random. MNAR data sets generally lead to biased estimates. Another version of MNAR would be a student who feels so unsafe at school that he or she doesn't complete the survey.

The mechanisms that give rise to the blanks in a data set must be understood before undertaking to "fill in the blanks" as it were. However, there are workarounds, and those are the subject of the remainder of the chapter.

Four Basic Approaches

In the previous chapter, We saw two ways to deal with missing observations: simply omitting a sparse variable (one with many missing observations) from the analysis or replacing missing values with zero. Little and Rubin (2002, pp. 19-20) propose a taxonomy of four categories of missing-data strategies.

1. **Procedures based on completely recorded units**. If there are relatively few missing observations, use casewise deletion and forge ahead. Similarly, if a variable has mostly missing values, omit it from the analysis.

2. **Weighting procedures**. This is common with survey data where population subgroups appear in the sample disproportionately to their occurrence in the population. Many platforms in JMP anticipate the use of sampling weights, which are often supplied with public-use survey data. We'll demonstrate this shortly using the YRBS data. The creation of sampling weights is beyond the scope of this discussion. Readers who want to create sampling weights for their own work can consult *Using JMP* for a discussion of stratifying a random subset.

3. **Imputation-based procedures**. There are several methods described below that fill in blank cells with plausible estimates, and then the analyst applies methods that require complete cases. JMP offers some of the available methods for imputation.

4. **Model-based procedures**. These approaches involve modeling using the observed data, applying maximum likelihood methods for estimation of parameters and variability.

In the remaining sections, we'll take up examples using the first three categories, leaving model-based procedures for the next chapter. As noted earlier, the focus here is largely on ways of wrangling the available data to prepare it for traditional analysis, so most of what follows focuses on imputation for missing data.

Working with Complete Cases

The simplest approach to missing data is to omit cases (rows) with blanks. This is an acceptable strategy if the missing cases are rare and if you can reasonably assume that cases are missing completely at random. Many analysis platforms drop cases with missing data automatically, but if it suits that analyst's purposes, you might select rows with missing data in a column (**Rows ▶ Row Selection ▶ Select Where**) and then **Hide/Exclude** those rows. There really is not much more to know.

Recall from the previous chapter that sometimes an entire column is empty or nearly so. Such variables will not contribute much information to a model, and might be deleted, omitted from a subset of the data table, or just not selected as potential explanatory factors in a multivariate model.

Analysis with Sampling Weights

Public-use survey data collected via complex sampling designs generally supply sampling weights in the data table. At the time of data collection, the database creators are generally aware of the proportion of respondents from various population strata. For example, in the YRBS data introduced early, the researchers know how many students are in each grade level nationally.

They also know how many respondents are in each grade level. From these known figures, they can compute weights to compensate for the proportional misrepresentation across strata.

When the sampling weights are available, we can accommodate the "missing" students with no further data wrangling. That issue was addressed when the weights were computed, and we just need to identify the weighting column within the applicable analysis platform. To illustrate, we'll use the **Health Risk Survey** table from the JMP sample data. Open it now.

In the Columns panel, click the disclosure triangle next to Hidden Columns (6/0) to display the columns that contain the sampling weights, primary sampling units (PSUs), and other background variables.

Figure 10.1: Sampling Weights and Other Background Variables

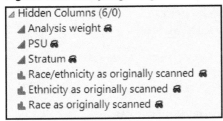

As shown in Figure 10.1, the **Analysis weight** column is initially hidden in the data grid. **Unhide** it now, and note that every individual in the sample has a weight value between 0 and 10, with a mean value of 1.0, which can be confirmed by examining the distribution of the **Analysis weight** column (not shown here). Intuitively, think of the weights as compensating for youth segments that are over- or under-represented in the sample. A teen with an analysis weight greater than one is representing a relatively large number of peers missing from the sample.

To understand the impact of sampling weights, let's look at a simple bivariate analysis using one pair of variables from the survey. Because there are so many columns in this table, the table creator has defined two more column groupings in addition to the hidden columns group. Click the disclosure triangle next to **Multiple Choice Questions (94/0)**, where we will find our two variables.

1. Select **Analyze ▶ Fit Y by X**. Cast **How much do you weigh** into the **Y**, **Response** role and **How tall are you** into the **X**, **Regressor** role.
2. Click **OK**.
3. In the bivariate fit report, click the red triangle next in the upper left, and choose **Fit Line**.
4. Next, in the bivariate fit report, click the red triangle once again, and select **Redo ▶ Relaunch Analysis**. This time, cast Analysis weight into the Weight role and click OK.

Figure 10.2 shows the linear regression output for both analyses. The unweighted analysis is on the left. Notice the differences in the two reports, beginning with the difference in sample size. The original sample has 13,119 complete cases out of 13,917 total cases because we have height and weight measurements for 13,119 of the respondents. After weighting, the approximate adjusted sample size becomes approximately 13,182. The weighting does not address the heights and weights of those respondents who didn't provide answers to the two questions, but it does compensate for their peers who were not present in the sample at all.

Most of the other reported statistics are different: The coefficient estimates have changed, the standard errors have become smaller, and the fit is superior. All significance-related results have improved as a direct result of incorporating the analysis weights.

Figure 10.2: Linear Regression with and without Sampling Weights

Linear Fit

How much do you weigh = -75.35129 + 84.714797*How tall are you

Summary of Fit

RSquare	0.267915
RSquare Adj	0.267859
Root Mean Square Error	14.44243
Mean of Response	68.19234
Observations (or Sum Wgts)	13119

▷ **Lack Of Fit**

Analysis of Variance

Source	DF	Sum of Squares	Mean Square	F Ratio
Model	1	1001268.5	1001269	4800.316
Error	13117	2735994.6	209	Prob > F
C. Total	13118	3737263.2		<.0001*

Parameter Estimates

Term	Estimate	Std Error	t Ratio	Prob>|t|
Intercept	-75.35129	2.075639	-36.30	<.0001*
How tall are you	84.714797	1.222712	69.28	<.0001*

Linear Fit

How much do you weigh = -79.72711 + 86.88597*How tall are you

Summary of Fit

RSquare	0.282131
RSquare Adj	0.282077
Root Mean Square Error	14.23027
Mean of Response	68.13355
Observations (or Sum Wgts)	13182.69

▷ **Lack Of Fit**

Analysis of Variance

Source	DF	Sum of Squares	Mean Square	F Ratio
Model	1	1043921.3	1043921	5155.148
Error	13117	2656202.0	203	Prob > F
C. Total	13118	3700123.3		<.0001*

Parameter Estimates

Term	Estimate	Std Error	t Ratio	Prob>|t|
Intercept	-79.72711	2.063085	-38.64	<.0001*
How tall are you	86.88597	1.210122	71.80	<.0001*

Imputation-based Methods

When there are too many missing cells to ignore, it might be possible to use educated estimates of those values so that we have complete cases to work with. The methods used to construct plausible replacement values are known collectively as *imputation methods*. In this section, we look at examples of these approaches to "filling in the blanks."

A word of caution is important here: JMP will make the requisite calculations and replacements on request, but will not make judgments about whether a particular method is logically, theoretically, or practically justified in a given study. That responsibility belongs to the analyst, who must have sufficient domain knowledge to make the choice. Imputation has substantial consequences for model-building and analysis, and should be approached with eyes wide open.

In this section, we return to the example of the Summer Olympic medals data (**Olympics Query.jmp**), and begin by considering those participating countries that won no medals at all in a given year. The first two approaches that we'll consider are quite simple: Either replace the empty cells with 0s—that is, these countries earned no medals in a given year—or code the blanks as a special informative case.

Recode

Recall that in the original data table, we had medal data only about those countries whose teams actually won medals in a particular edition of the summer games. A country that sent a team but did not finish in the top three in any event appears in the original medals table with blanks in the medals column. The quickest way to dispatch this issue is with **Cols ▶ Recode** (now in the **Cols** menu in JMP13). Just replace the missing values with 0.

In this approach, we uniformly replace all missing values with a constant. Given the way that the data were assembled for this study, this is an appropriate action. The countries were in the games, but came home with no medals. On the other hand, if we had widened the data gathering to include all nations of the world, including those that do not participate in the Olympics movement, this approach would not be suitable. We would want to distinguish between non-participants and non-winners, and treating all blanks as zeros would bias further analyses. Let's do that now, as illustrated in Figure 10.3.

1. Select the column **TotMedals**.
2. Select **Cols ▶ Recode**. In the top row, enter a **0** in the first cell under **New Values**.
3. Click **Done**, which opens a menu of options. Choose **New column**.
4. Save this table as **Olympics Query 2**.

Figure 10.3: Recoding Missing Values as a Constant

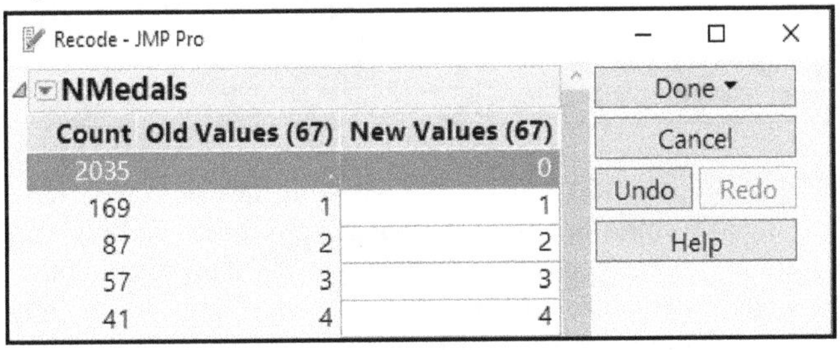

This creates a 27th column in the table. Whereas **TotMedals** had 481 complete rows and 2002 missing values, the new **TotMedals2** column has 2,483 complete cases.

Informative Missing

In some studies, the very fact that some cells are blank carries meaning. As Sall (2013) notes, "when you are studying the data to predict if a loan is bad and the loan applicant leaves his income or his current debts missing, we are probably better to assume that the applicant is being evasive here for a reason." This would be a case of MNAR—missing not at random. Rather than omit the case or attempt to estimate a plausible income value, we might create a code to indicate that a value is missing and include the case within the analysis.

In the current study, consider countries that do not collect or report figures on the proportion of girls enrolled in primary and secondary education. These countries might not place a high priority on the education of female students, or might not have the statistical infrastructure to gather such data. For the sake of illustration, we can investigate the notion that these data are not missing at random.

To demonstrate, we'll look at the relationship between the two gender equity variables in the table, using the primary enrollment ratio to predict the secondary ratio. Initially, we fit a quadratic model.

1. Select **Analyze ▶ Fit Y by X**. Cast **Gross enrollment ratio, secondary, gender parity index (GPI)** as Y and the corresponding primary ratio as X, and click **OK**.

2. Now fit a 2nd-degree polynomial (quadratic) model, with the results as shown in Figure 10.4.

Figure 10.4: Initial Quadratic Model

◢ **Summary of Fit**

RSquare	0.739242
RSquare Adj	0.738754
Root Mean Square Error	0.130133
Mean of Response	0.887348
Observations (or Sum Wgts)	1072

▷ **Lack Of Fit**

◢ **Analysis of Variance**

Source	DF	Sum of Squares	Mean Square	F Ratio
Model	2	51.321734	25.6609	1515.296
Error	1069	18.103043	0.0169	**Prob > F**
C. Total	1071	69.424777		<.0001*

◢ **Parameter Estimates**

Term	Estimate	Std Error	t Ratio	Prob>|t|
Intercept	-0.521878	0.033758	-15.46	<.0001*
Gross enrolment ratio, primary, gender parity index (GPI)	1.5286404	0.034874	43.83	<.0001*
(Gross enrolment ratio, primary, gender parity index (GPI)-0.9062)^2	0.4247341	0.100185	4.24	<.0001*

This model accounts for 74% of the variation in Y, and has considerable room for improvement. To see how the info-missing approach changes the result, we'll next refit the model, but this time treat missingness in X as informative. We'll do so using the **Fit Model** platform, and designate the primary ratio as informative missing. This creates a pair of variables: One is just a dummy indicator for blank cells. The other builds a formula equal to the observed value when available, and the other is the mean of the nonmissing values.

3. Select **Analyze ▶ Fit Model**. Select the secondary parity ratio as **Y** once again.

4. Right-click **Gross enrollment ratio, primary, gender parity index (GPI)** and select **Distributional ▶ Informative Missing**. (See Figure 10.5.) Notice two new virtual columns selected at the bottom of the Select Columns panel.

Figure 10.5: Tag a Column Informative Missing

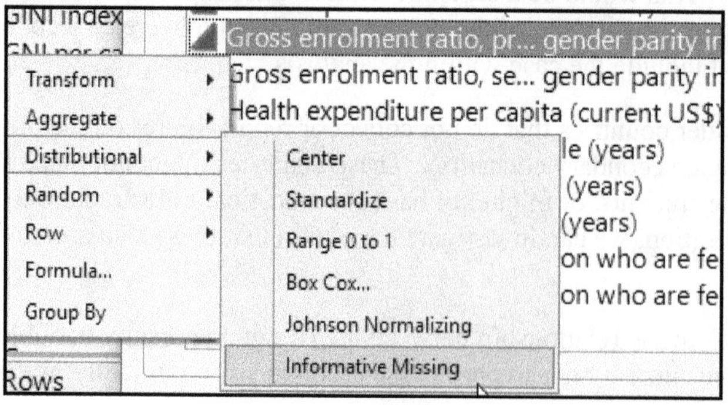

5. With the new columns selected, click **Macros**, and choose **Polynomial to Degree**, using the default **Degree** setting of 2. ThIS adds three model effects, corresponding to the **Is Missing** dummy, X, and X^2.

6. Click **Run**.

In the regression plot at the top of the fit model report, we now see two parallel quadratic curves, corresponding to the two different values of the dummy variable. Further exploration of the statistical results (Figure 10.6) shows that N has increased by bringing in 38 more cases (missing values of Y are still not used), and the overall goodness of fit has deteriorated and RMSE has increased.

Figure 10.6: Model Results Using Informative Missing

⊿ Summary of Fit					
RSquare	0.671921				
RSquare Adj	0.671029				
Root Mean Square Error	0.150923				
Mean of Response	0.889683				
Observations (or Sum Wgts)	1108				

⊿ Analysis of Variance					
Source	DF	Sum of Squares	Mean Square	F Ratio	
Model	3	51.501659	17.1672	753.6802	
Error	1104	25.146755	0.0228	Prob > F	
C. Total	1107	76.648414		<.0001*	

⊿ Parameter Estimates				
Term	Estimate	Std Error	t Ratio	Prob>|t|
Intercept	-0.528758	0.040487	-13.06	<.0001*
Informative[Gross enrolment ratio, primary, gender parity index (GPI)]	1.536196	0.041848	36.71	<.0001*
(Informative[Gross enrolment ratio, primary, gender parity index (GPI)]-0.9151)*(Informative[Gross enrolment ratio, primary, gender parity index (GPI)]-0.9151)	0.4247341	0.116191	3.66	0.0003*
Is Missing[Gross enrolment ratio, primary, gender parity index (GPI)][1-0]	0.0958416	0.025745	3.72	0.0002*

Parameter estimates have changed slightly, and the significance levels are not as compelling as in the first model. The point here is that informative missingness is one approach to consider, and it can impact the magnitude of coefficients as well as the standard errors in a model. In this case, the approach has only a modest effect on the number of usable cases, and results in a weaker model fit.

Multivariate Normal Imputation

In Chapter 9, we saw the **Explore Missing** platform and illustrated how it helped identify missing values and patterns across columns. (Please refer back to Figures 9.12 and 9.13.) The **Missing Values** report provides two methods of imputing missing data, called **Multivariate Normal Imputation** and **Multivariate SVD Imputation**. This section explains the former method, which is identical to the approach taken when selecting **Impute Missing Data** from the **Multivariate** platform.

For this example, we will start with the multivariate platform approach.

1. Select **Analyze ▶ Multivariate Methods ▶ Multivariate**. Select the nine wealth-related columns.

2. After the report opens, click the red triangle next to **Multivariate**, and choose **Save Imputed Data.** This automatically creates and saves new columns with the imputed data in the original data table.

Invoking the imputation is simple: You click the button. What does this method do, and when might you best use it? Let's return to the example of Chapter 9. In the Olympics data table, we have nine continuous variables addressing wealth and its distribution in each country for each year. The first of the nine (GDP per capita) is most nearly compete, with only 655 observations missing from the 2,483 rows. However, we are missing 1,368 of the GNI per capita observations.

We selected all nine wealth columns to create the **Missing Values** report. At that point, the algorithm depends on the *estimation method* selected when launching the **Multivariate** platform. Computations of correlations, for example, depend on the method chosen.

Figure 10.7: Available Estimation Methods

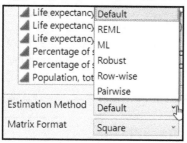

As show in Figure 10.7, we can choose among six estimation methods. For this illustration, we choose default and in turn JMP selects on of three methods depending on characteristics of the data table:

- Rowwise estimation if there are no missing cells in the data table.
- Pairwise estimation if there are missing values and either the data table has more than 10 columns or more than 5,000 rows, or if the number of columns exceeds the number of rows.
- REML (Restricted Maximum Likelihood) estimation is selected otherwise by JMP, and will be applied in this case. A full explanation of the method is beyond the scope of this book, but understand that it is helpful for compensating for bias when estimating variances and covariances (and hence correlations), particularly in the presence of missing data.

Users who want to investigate other options should consult the JMP book *Multivariate Methods*, found under **Books** on the **Help** menu.

The imputed values for blank cells in a column are expected values for the mean, conditioned on the nonmissing values in each row. The computational formula uses the mean and covariance matrix, and the resulting data set will be complete, which is to say there are no missing values.

Please note that imputation *replaces the missing values with computed values in their original locations.* Hence, once you issue the command, the original data table in memory changes. If you want to be able to work with both the original data and the imputed data, use **Save As** to preserve the altered data table while preserving the original.

In this example, where wealth indicators for each nation often increase over time, this method can lead to implausible values. Consider, for example, the case of the first nation in the table, Afghanistan. Figure 10.8 shows a simple overlay plot for the imputed GDP per capita in constant US dollars.

Figure 10. 8: Observed and Imputed GDP per Capita in Afghanistan

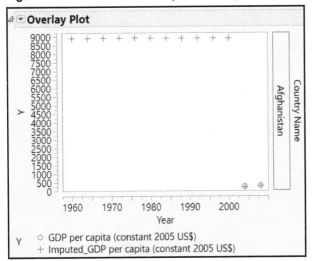

Of the 13 years, we had observed measurements for only 2, as indicated by the red circles in the lower right. Based on the available data in the other 8 columns for Afghanistan each year, the method computes an expected value. For the 11 earlier years, the imputed values are absurdly high and constant.

Multivariate SVD Imputation

When a data table has more columns than rows—or in any event, a large number of columns—it might be advisable to use an imputation method that relies on *singular value decomposition (SVD)*. SVD basically transforms the initial data matrix as a series of rotations and rescalings. Readers interested in the underlying mathematical detail should consult Appendix A in the JMP Book *Multivariate Methods*, found in the **Help** menu.

Here we illustrate the method by starting anew with the Olympics data, having just generated a **Missing Values** report for the same nine wealth-related columns. This time, however, we select **Multivariate SVD Imputation**.

This selection opens the dialog shown in Figure 10.9. JMP recommends settings, which we'll use in this example. Click **OK** in the dialog.

Figure 10.9: Settings for SVD Imputation

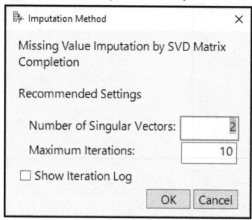

After you accept the recommended settings, an **Alert** pops open with an audible alarm (Figure 10.11) to remind the user that the data table's missing cells have now been replaced by imputed values. Once again, click **OK** and note that the Imputation Report indicates the number of cells replaced.

Figure 10.10: JMP Alert Regarding Imputation Results

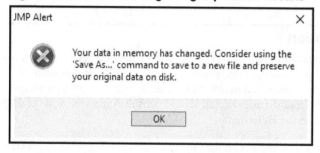

As we did in the prior section, we can visually inspect some of the imputed values as a reasonableness check, and as a way of beginning to understand the differences in different methods. Figure 10.11 displays line plot comparable to the one shown in Figure 10.8.

Figure 10.11: GDP Per Capita, Afghanistan, with Observed and Imputed Values

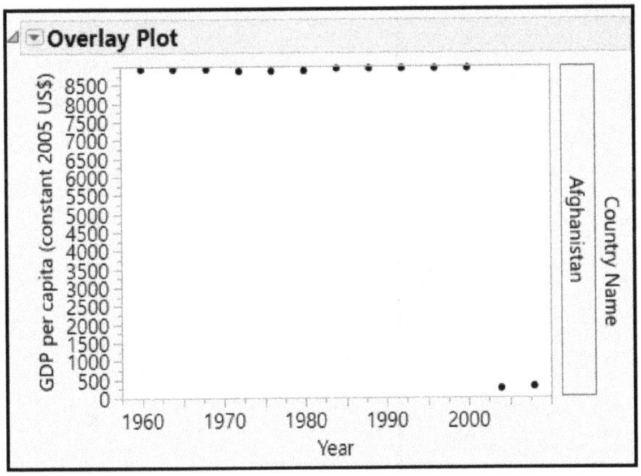

Much as before, the imputed values (all but the final two observations) are unreasonably large and we are no closer to the observed data using the SVD method than the normal method. In this instance, the data under examination is not particularly wide (just 9 columns in comparison to 2,483 rows), so the SVD method is not particularly suitable for this data set.

Special Considerations for Time Series

In this example, we really have panel data: a relatively stable set of 193 nations observed 14 times at regular 4-year intervals. For many of the countries in the data set, some of the economic time series are left-censored. That is, these countries did not report the statistics in the earlier years under study. We've seen that Afghanistan's data started in 2004, and the first 11 observations are missing. Given the political turmoil in Afghanistan for much of its recent history, it is likely to be difficult to impute plausible GDP values.

However, in countries with few missing values and stable rates of GDP and population growth, we might apply a simple time series forecasting technique to impute the missing observations. Though it is not at all clear how you might reliably do this at scale with this data set, recognizing that the time period of the study is so long and so marked by war and political upheaval, you might find this strategy useful in another setting.

Take, for example, per capita GDP in Bhutan in the pre-imputation raw data, depicted in Figure 10.12. We have 9 observations, beginning in 1980. Multivariate normal imputation estimated a constant GDP of more than $9,100 per capita for each of the first five missing years, which is not remotely close to the observed data.

Figure 10.12: GDP Growth in Bhutan

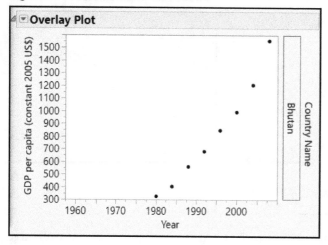

Consider the following approach:

1. In the original data set, we select just the 14 rows for Bhutan.
2. Choose a suitable linear or time series method to impute the early years.
   a. In this example, a model using a log transformation and **Fit Y by X** would obviate the need for a reverse chronology.
   b. In other examples, one of the time series methods might be superior. To use those methods, you would first need to create a new column equal to the (2012 – the numeric value of **Year**). This has the effect of reversing the chronological order of the data.

If we go ahead and fit a model of log(GDP) versus Year to this set of data (first converting the ordinal variable **Year** to continuous), and then save predictions, we generate plausible estimates for the first five unobserved years. Figure 10.13 shows the model fit as well as the imputed values, which appear at triangles.

Figure 10.13: Log-Linear Imputation for Smooth Time Series

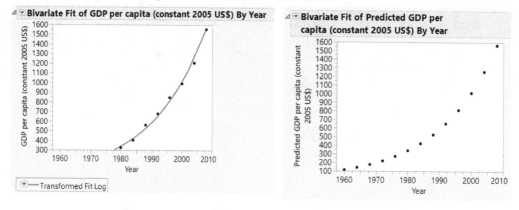

Conclusion and a Note of Caution

Missing data presents numerous challenges, and is among the most common data preparation hurdles. This chapter has presented several useful approaches that take advantage of JMP functionality and has also introduced a small amount of theory to guide practitioners. Following this section is a long list of references on the theory, as well as some examples from various disciplines.

As a general rule, statisticians tend to be reluctant to discard data, and many modeling methods require complete cases for analysis. When it is possible and responsible to use available data to estimate missing data, that is a good thing.

As noted earlier, though, it's a risky matter to replace unknowns with estimates. Imputation of missing data calls for a blend of domain knowledge and methodological understanding. As we've seen, JMP—or any software for that matter—will rapidly follow a user's instruction to impute huge numbers of observations. Use these methods with eyes wide open.

The next chapter takes up the matter of data transformations: methods to cope with outliers, non-normality, and non-linearity. In addition, we discuss a method that can also help with missing data, known as *Principal Components Analysis*.

References

Brady, Brady. 2014. "Imputation Addin." JMP User Community. Downloaded from https://community.jmp.com/t5/JMP-Add-Ins/Imputation-Addin/ta-p/22555.

Centers for Disease Control and Prevention. 2016. *2015 YRBS Data User's Guide.* (June), Atlanta, GA: CDC.

Gelman, Andrew, and Jennifer Hill. 2006. "Missing-data imputation." *Data Analysis Using Regression and Multilevel/Hierarchical Models*, Chapter 25. New York: Cambridge Univ. Press. Available at http://www.stat.columbia.edu/~gelman/arm/missing.pdf.

Horton, Nicholas. J., and Ken P. Kleinman. 2007. "Much Ado about Nothing: A Comparison of Missing Data Methods and Software to Fit Incomplete Data Regression Models." *The American Statistician*, Vol. 61, No. 1 (Feb.), pp. 79-90.

Little, Roderick J. A., and Donald B. Rubin. 2002. *Statistical Analysis with Missing Data,* Second Edition. Hoboken, NJ: Wiley.

McCormack, Don. 2015. "It's a Dirty Job, but Someone Has to Do It." Presentation at *JMP Discovery Summit 2015*. Downloaded from https://community.jmp.com/t5/Discovery-Summit-2015/It-s-a-Dirty-Job-but-Someone-Has-to-Do-It/ta-p/23806.

Penn, David A. 2007. "Estimating Missing Values from the General Social Survey: An Application of Multiple Imputation. *Social Science Quarterly*, Vol 88, No. 2 (June), pp. 573-584.

Rubin, Donald B. 1996 "Multiple Imputation after 18+ years." *Journal of the American Statistical Association,* Vol. 91, No. 434, (June) pp. 473-489.

Sall, John. 2013. "It's not just what you say, but what you don't say: Informative missing values." *JMP Blog*, October 29. Available at https://community.jmp.com/t5/JMP-Blog/It-s-not-just-what-you-say-but-what-you-don-t-say-Informative/ba-p/30328.

SAS Institute Inc. 2016. "Explore missing values utility", in JMP 13 *Basic Analysis*, Cary NC: SAS Institute Inc.

SAS Institute Inc. 2016. JMP 13 *Multivariate Methods*, Chapter. 3. Cary NC: SAS Institute Inc.

Sterne, Jonathan A. C., and Ian R. White, John B. Carlin, Michael Spratt, Patrick Royston, Michael G. Kenward, Angela M. Wood, James R. Carpenter. 2009. "Multiple imputation for missing data in epidemiological and clinical research: potential and pitfalls." *British Medical Journal*, Vol. 339, No. 7713 (18 July), pp. 157-160.

Svolba, Gerhard. 2006. *Data Preparation for Analytics Using SAS.* Cary NC: SAS Institute Inc.

Chapter 11: Data Preparation for Analysis

Introduction

Once we have all of the relevant data in a single JMP data table, there might still be a need for "preprocessing" prior to analysis. We treat the topics in this chapter after the chapter on missing data, because some operations demonstrated here will proceed more smoothly if we have resolved problems of missing data. On the other hand, the technique of Principal Components Analysis (PCA) is itself a powerful tool for overcoming missing data challenges.

Data reduction is one class of methods that we consider in this chapter, which might initially strike readers as odd. In many studies, analysts struggle with the paucity of data. But with ever-growing data sources and automated data generation, it is increasingly important to confront the challenges of having too much data—enormous data sets can burden computing capacity and make interpretation more difficult.

One theme running through this chapter is that of marginal information gain. In the case of *high-dimensional* data sets—that is, those with large number of variables (columns), you might sensibly ask "After building a model that accounts for k variables, how much incremental information do we gain from variable $k + 1$?" Similarly, if we have a table of 100,000 rows, how much insight would come from an additional 1,000 rows? How much information would we lose by modeling based on a sample half the original size? As we introduce and review strategies in the pages that follow, keep this theme in mind.

Another idea to keep in mind is that each of the examples should not be considered comprehensive, but rather suggestive and illustrative. Each of the techniques can be further customized, and readers should consult the JMP documentation to learn about more options. Every study is different, and the needs for preprocessing will depend on the nature of the analytic goals, the substantive knowledge of the analyst, and the characteristics of the data.

Common Issues and Appropriate Strategies

The key drivers in deciding how to transform the data in a table are characteristics of the data (outliers, skewness, missingness, high-dimensionality, and so on) and the plan for analyzing the data. For example, skewness is problematic for traditional inference methods based on assumptions of normality, but it is not necessarily troublesome with other techniques.

Borrowing from Han, Kamber, and Pei (2011, Chapter 3), we can define four general types of data characteristics that lead us to transform a data table; these are summarized in Table 11.1 below. Additional strategies exist, but the chapter covers a range sufficient to get a foothold for most problems. The rest of the chapter covers each problem area in sequence.

Table 11.1: Common Issues Addressed by Transformation

Category	Common Problem	Common Solutions
Distribution of observations		
	Noisy data: high degree of variation in continuous data	Binning, nominal discretization, smoothing (for time series)
	Skewness or outliers	Functional transformation (for example, logarithmic)
	Scale differences among variables to be used	Normalization
	Too many levels in a categorical variable	Recoding, clustering
High Dimensionality: Abundance of Columns		
	Correlated or redundant variables	Drop columns, Principal Components Analysis

Category	Common Problem	Common Solutions
	Sparsity (many 0s) or missing observations across columns	Principal Components Analysis
Abundance of Rows		
	Table partitioning for training, validation, and testing	Create validation column, subset
	Variety of observational units	Aggregation by strata, cluster, or time period
Date and Time-Related Issues		
	Inconsistent date or time data	Date functions

Distribution of Observations

The first category concerns situations in which we want to tame either the shape or the variability of one or more variables. We'll consider four cases, three of which involve continuous data and one involving categorical data.

Noisy Data

By noisy, we refer to a situation in which a variable has a high degree of random variation, or where the analytic purpose is somehow impeded by the precision of the recorded observations. The fundamental idea here is to smooth the data in a way to reduce the noise of the variability, revealing the essential signal in the data without substantial information loss.

One common approach, sometimes known as *binning*, is to gather similar observations into *bins*, according to their location on the number line, and to replace the observed values with binned values. To see how this works, let's return to our Olympic medals data set (**Olympics Query.jmp**), and focus on life expectancies in different nations. For the purposes of this example, we'll assume that our modeling plan will use life expectancy as a model feature (independent variable), and also that we'll want to estimate the impact of "large" differences in life expectancy, rather than the marginal impact of (for example) one additional year of life expectancy

1. Select **Analyze ▶ Distribution**. Choose **Life expectancy at birth, total (years)** leaving all default settings and click **OK**. This produces the report shown in Figure 11.1.

Figure 11.1: Initial "Noisy" Distribution

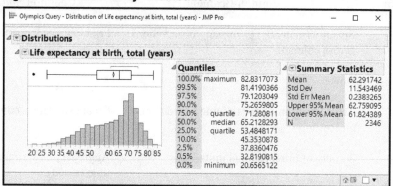

The entire data table consists of 2,483 rows, and this column shows observations for 2,346 of them, so we have a small amount of missing data. The distribution is skewed, with just a handful of outliers. However, if our goal is to model the number of medals won in a year, we might not need the granularity of observations recorded here.

How might we establish bins or groupings of this variable to transform it from 2,346 unique values into perhaps ten bins? Several strategies present themselves, each of which has numerous variations on a single theme. Readers are encouraged to explore JMP documentation further after taking these initial steps.

2. Modify the histogram to create the desired number of categories, and then create a new variable representing either the bin number (for example, 1 through 10) or the mean of the bin location. Either approach will yield a fixed number of bins containing different numbers of observations.

3. Create *k* equally sized groupings of an approximately equal number of observations based on quantiles or ranks.

4. Because the values conveniently range from the 20s to the 80s, you might even simply round values to the nearest 10s, building a formula with the **ROUND** function.

> With time series data, you might apply one of several smoothing methods such as moving averages, ARIMA, or exponential smoothing, or consider taking first differences. Later in this chapter, we take up special issues related to time- or date-stamped observations. A full treatment of time series methods is beyond the scope of this chapter, but interested readers should consult the "Time Series" chapter in JMP 13 *Predictive and Specialized Modeling* (SAS 2016).

We'll illustrate the first two options here, both of which use **Histogram Options** under the red triangle next to the column title in the **Distributions** report.

5. Click the red triangle next to **Life expectancy at birth, total (years)**, and select **Histogram Options ▶ Set Bin Width**. This opens a small dialog with one entry (not shown) asking you to specify a **New Bin Width**. With this particular range of data, a bin width of 7 establishes 9 bins.

NOTE: Experienced users also know that we can modify the number of bins interactively by sliding the left- or right-end of the horizontal axis. Use whichever technique suits your work flow.

6. Again, click the red triangle, and click **Save**. This opens an additional menu of options (Figure 11.2), the first four of which are relevant to creating bins of equal width. Missing values are skipped.

   ○ **Level Numbers** refers to the consecutive bin number in which the original value falls. The numbers start at 1 for the lowest-valued observations, and the new variable has modeling type nominal.

   ○ **Level Midpoints** refers to the mean value between the lower and upper bounds of the bin. For example, if there were a bin ranging from 20 to less than or equal to 30, the midpoint would be 25.

Figure 11. 2: Save Options for the Distribution Report

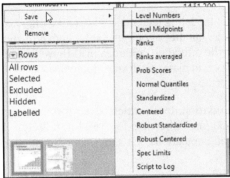

7. For illustration, choose **Level Midpoints**.

   Now look in the data table. There is a new final column called Midpoint Life Expectancy, total (years). In Figure 11.3, we see the effect of this transformation.

Figure 11.3: Comparing Raw Data and Discretized Values

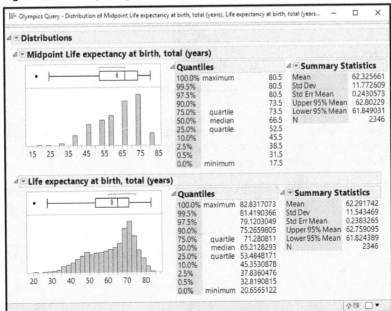

8. Repeat the prior steps, but this time save the **Level Numbers**. The resulting histogram of the discretized data has the same form as the midpoint transformation, with the same number of observations within each bin, but we lose information years of life expectancy, and the new column is nominal. In some contexts, it might be desirable to attach meaningful descriptors to the new categories, such as "extremely low" to "extremely high."

Binning can sometimes offer a resolution to the issue of outliers. As with so many other issues, the appropriateness depends on the analytic purposes and the knowledge of the analyst. When appropriate, you might accumulate outlying values in the lowest or highest bin. For example, age data often have highest categories akin to 80 years and older.

Skewness or Outliers

If the analysis plan includes the use of a modeling technique with a normality assumption, highly skewed distributions or those with a large number of outlying values are problematic. A common strategy is to apply a transformation to the offending columns. In JMP, we have the choice of creating a new column using a formula or, in many platforms, doing a virtual transformation within the context of the analysis platform.

Consider, for example, the column **Population, total**, which is very strongly right-skewed with numerous high-end outliers. The distribution is shown in Figure 11.4. It is clear that most country-year observations have moderately small populations, but some have exceptionally large values.

Figure 11.4: Distribution of Population

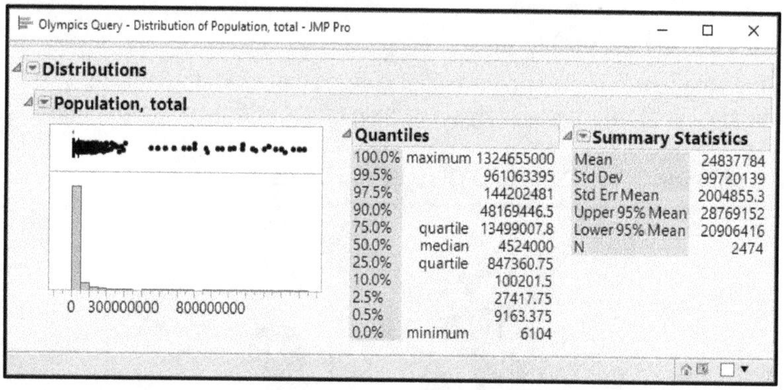

Perhaps the most widely used transformation is logarithmic, which tends to pull very high values in a distribution to the left, while having relatively little effect on lower values. Experienced JMP users know that we can create a new column using the Formula Editor to add another variable to the table equal to the logarithm of population.

If you prefer not to enlarge the data table with another column, most analysis platforms now allow for this approach.

1. Return to the **Analysis ▶ Distribution** platform.
2. Position the pointer over the column title **Population, total** in the list of columns and right-click.

3. Select **Transform ▶** and then **Log**.

    This adds a new italicized item to the column list: **Log[Population, total]**. The italics indicate that the transformation is temporary, and available for the current analysis. Notice that the new column does not appear in the data table.

4. Select this new item as the **Y, Columns** role and click **OK**.

The result of the transformation is shown in Figure 11.5. The transformed data is far more symmetric and more nearly normal in shape than the original data in Figure 11. 4. Also, the number of outlying values has been reduced. In fact, a quantile range outlier analysis of the type illustrated in Chapter 9 (but not shown here) indicates 46 extreme values in the original data, but none at all in the transformed data.

Figure 11.5: Distribution of the Log of Population

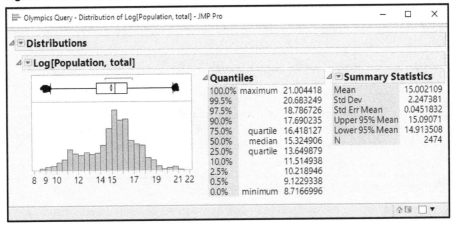

Note that logarithm completely changes the scale of the observed data. In the original data, the minimum value of 6,104 refers to persons (the 1960 population of the tiny South Pacific island nation of Tuvalu). The minimum log value still refers to Tuvalu in 1960, but it is the log of population. When interpreting modeling results, the analyst needs to exercise care about the change of scale.

There are several transformation functions available in addition to natural log, and these are listed in Figure 11.6. Logs are very helpful for distributions that are strongly right-skewed and for straightening non-linear relationships between variables. Consult the JMP documentation for full details about the other functions.

Figure 11.6: Available Transformation Functions

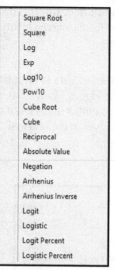

An alternative method for modeling in the presence of outliers is to use robust methods, available in many platforms. Because that approach is more a part of modeling than of data preparation, it is not treated here—but keep it in mind as yet another option.

Scale Differences among Model Variables

For some modeling methods, the comparative scale of model variables is an important consideration. For example, distance-based methods like k-means clustering or k-nearest neighbors aim to gather observations that are proximate in multidimensional space. When applying such techniques, it is often advisable to express variables in a standard scale like z-scores.

We can do this in much the same way as functional transformations. To continue our example, we might have a model that includes life expectancy and population, among other covariates. If we are seeking clusters of countries with similar life expectancy and populations, it might be useful to express the variables in standardize form. We would do this as follows, using the distribution platform once again as the exemplar.

1. Select **Analyze ▶ Distribution**. First, highlight both **Life expectancy at birth, total (years)** and **Population, total**, and cast them as **Y**.
2. With both columns still selected, position the pointer over either of them and right-click.
3. On the menu, select **Distributional ▶ Standardize**.
4. As before, this appends two new virtual columns. Cast them at **Y** as well.

The results are shown in Figure 11.7. The noteworthy features of this report are that the raw and standardized histograms are identical, but the mean and standard deviation of the two standardized variables are 0 and 1 respectively.

Figure 11.7: Effect of Standardizing Columns

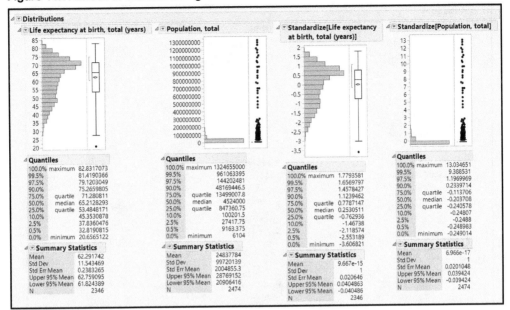

Too Many Levels of a Categorical Variable

Excessive variability is not restricted to continuous or discrete data. Categorical variables, whether nominal or ordinal, sometimes have so many categories as to muddy further analysis. When combining data from different sources, it might be necessary to reconcile inconsistent labeling or coding of categories. For such circumstances, the Recode facility can help.

We used **Cols ▶ Recode** earlier to replace missing values of Olympic medals won with zeros. In this section, we'll demonstrate two additional uses of the command with categorical data. Because the Olympics table has relatively few categorical columns, we'll use one of the JMP Sample Data tables.

1. Select **Help ▶ Sample Data Library**. Open the data table **Hollywood Movies**. This is a list of 136 profitable films. We'll focus on three nominal columns: Lead Studio Name, Theme, and Genre.

2. Select **Analyze ▶ Distribution**. Select the three columns just mentioned and look at their distributions. (See Figure 11.8 for part of the report.)

Figure 11.8: Three Categorical Variables with Many Levels

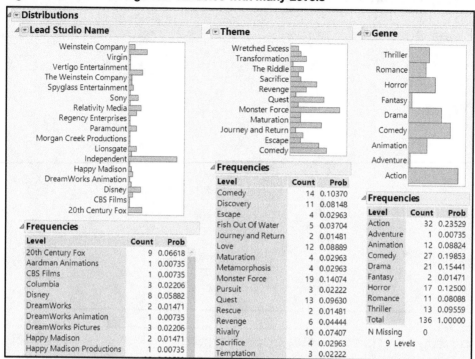

This data table lists 33 distinct studios, many of them appearing only once. What's more, there are some that seem to be listed by multiple names, and others that are related (like Pixar and Disney). Depending on the goals of a study, you might want to settle on a single name for the multiples and combine others under a parent company or simply "other" for the least-frequently appearing firms. The Recode command allows for all of these alternatives.

Similarly, for a particular study, 21 themes and nine genres might be excessive. We'll demonstrate Recode with the studio column. Reducing the levels of the others is an exercise left to the reader.

1. In the data table, select the **Lead Studio Name** column.
2. Select **Cols ▶ Recode**. Let's first combine rows with similar studio names.
3. With just a few cases to group, we could select the relevant **Old Values** (Figure 11.9), and right-click. This opens a pop-up that offers three options for a group name. For now, let's not select any of the options. There is a better way that we'll see shortly.

   Note that the grouped levels do not need to be sequential or contiguous. If we wanted to group Pixar and Disney, we'd just select the two values and right-click again.

Figure 11.9: Grouping a Few Categorical Levels

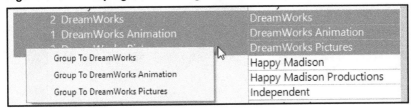

This approach is fine when there are a small number of changes to make. It does not scale well however. Recode can attempt to automate the process.

4. Click the red triangle at the upper left (Figure 11.10) and choose **Group Similar Values**. Accept all of the default options and click **OK**.

Figure 11.10: Recode Options for a Categorical Column

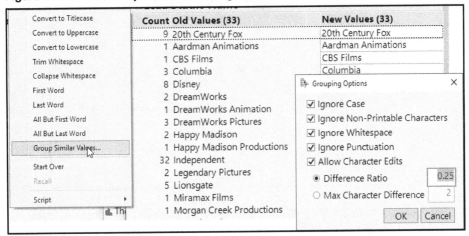

Using text analysis methods, JMP attempts to gauge the differences between pairs of character strings in the selected column. The methodology is explained in Chapter 4 of *Using JMP* 13 (2016) and we have two options for the comparison. Often, the default option works well. In this case, however, the difference ratio of 0.25 only finds two values to group: Weinstein Company and The Weinstein Company. If we adjust the ratio to 0.50, the results are better but not perfect as shown in Figure 11.11.

Figure 11.11: Automatic Suggested Groupings of the Lead Studio Names

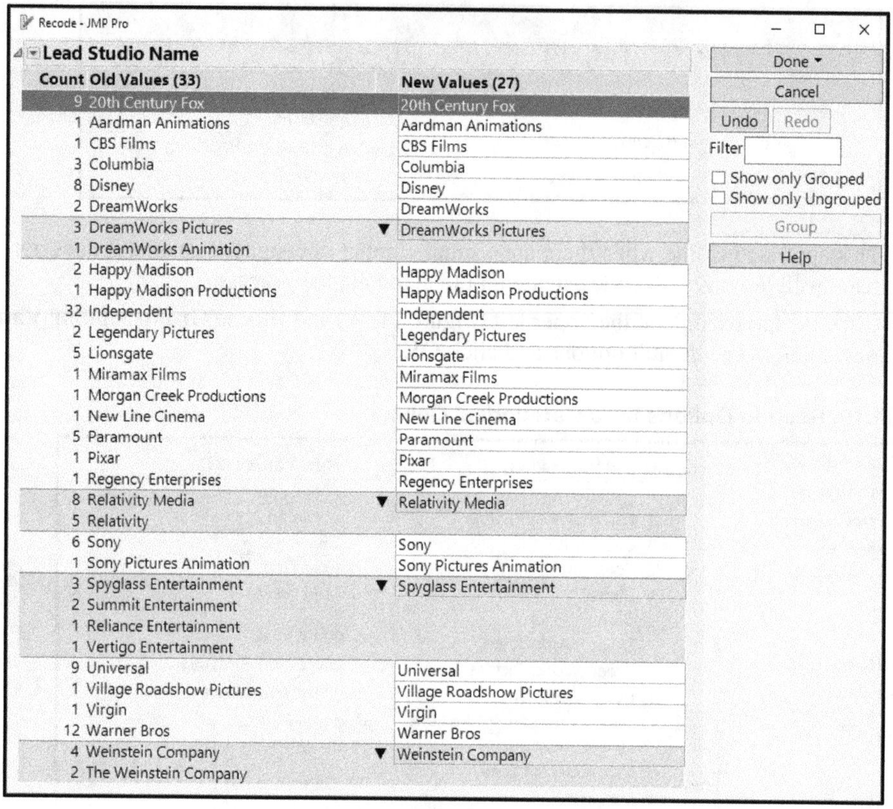

The method found four groups, but some (as they might say in Hollywood) are hits, and some are flops. Of the three Dreamworks entries, it combined two. It missed the two Happy Madison entries, but combined the Relativity Media and Relativity. The group beginning with Spyglass seems to be based on the presence of "Entertainment" in the four studio names. Further tinkering with the ratio parameter might lead to better results.

It is not necessary to accept all of the suggested groupings. Right-clicking the Entertainment group offers the option of ungrouping. To expand the Dreamworks group to include the omitted Dreamworks row, just highlight the three rows and form a group with the name of your choice.

When finished, click **Done**. Then opt to replace the values within the original column, create a new column of recoded values, or a new column containing a formula for the new values. If the original values were to change, the formula option dynamically updates the recoded values. JMP also offers the option of saving the Recode script to the data table, which is advisable for reproducibility. Saving and running the script recodes the data in place.

High Dimensionality: Abundance of Columns

Traditionally statistical analysts sought more data rather than less. In the current era, we are often confronted with more variables and more observations than needed. In this section and the one that follows, we consider a few strategies to overcome the embarrassment of riches that can arise from the plethora of data now available.

As noted in Chapter 9, particularly when working with automatically gathered sensor data, we might have data tables with very large numbers of columns that are irrelevant to the modeling task at hand, redundant with other columns in the table (collinear), and/or have unfortunate patterns of missing observations. We might compound the challenges by joining tables from various sources. Regardless of the data source, the "curse of dimensionality" can complicate the analyst's work. Thankfully, there are workarounds.

When columns are irrelevant to model-building, we can simply omit them from the model. Certainly, this type of variable de-selection is a standard part of statistical modeling. The issues of missingness, sparsity, and highly collinear variables are more challenging, and are discussed here.

Correlated or Redundant Variables

When building models that use multiple features, we ordinarily seek independent factors that contribute information that other factors do not contribute. When the information value of different features overlap, the marginal contribution of each additional model variable diminishes.

In some instances, the simplest solution to redundancy is to identify highly collinear columns and then only include the one that makes the highest contribution to model fit. Alternatively, if (for example) x_1 and x_2 are highly correlated, we might find that x_1 and a functional transformation of x_2 are less so. Hence, we transform the second variable and include both, thereby gleaning some of the information value of the additional variable.

In other instances, a combination of the original related columns can yield more information for a model than the original separate columns. One very useful method that can achieve this outcome is *principal components analysis* (PCA). PCA is a multivariate method that estimates and extracts a small number of underlying factors from a set of columns.

Conceptually, given a set of m variables whose combined total variation is V, PCA creates a set of m weighted linear combinations of the variables. The first principal component is that linear combination whose variance is the maximum of all possible combinations, capturing as much of the joint variability as possible. From there, the algorithm finds a second component that is orthogonal to the first (and therefore uncorrelated with the first) and that captures the maximum amount of remaining variation. This process continues through all m components.

In the simplest case, if there were just two correlated columns under consideration, the goal of PCA is to combine them in such a way that the first principal component captures most of their combined variability. And, therefore, we can build a model using either the first component rather than two variables, or we can use two uncorrelated principal components and (in the context of linear regression) thus bypass the issue of multicollinearity.

A very appealing characteristic of PCA is that it can create usable components even in the presence of missing data or columns with sparse data. Before demonstrating the technique, let's briefly review these two data problems.

Missing or Sparse Observations across Columns

Chapter 10 dealt extensively with the problems arising from blank cells within a data table. Sparsity refers to a different but comparable problem, in which we have a large number of distinct measures that happen to equal 0 over many rows. For example, text mining algorithms typically construct a term-document matrix showing the frequency with which specific words occur within a set of documents. Suppose that we have 1,000 documents containing a total of several thousand terms, and create a matrix with 1,000 columns and several thousand rows. Each cell of the matrix contains the frequency with which a given term occurs within a specific document. Most cells in such a matrix will contain frequencies of 0, and we refer to a matrix like this as *sparse*.

Analogously, we might consider a data array with customers in rows and products in columns, where cells contain the dollar value of a customer's purchases of each product. Here again, the matrix is likely to be sparse. The absence of purchase transactions yields zeros in many cells.

Collectively, the columns in a sparse matrix contain useful information, but individually any one column contains comparatively little information. PCA is one methodology to distill many columns into a relatively small number of model features.

A PCA Example

Let's continue with the Olympic medals analysis. Recall that we had several columns containing different measures of per capita wealth in the participating countries, with other columns referring to health (health-care spending and life expectancy). In Chapter 9, we saw that some of these columns were highly correlated with each other, and that many observations were missing.

To simplify this illustration, we'll confine ourselves to four of the wealth columns and two of the health columns. Before running a principal components analysis, let's look at the relationships among these six variables.

Still using our **Olympics Query.jmp** table as the active data table, do this:

1. Select **Analyze ▶ Multivariate Methods ▶ Multivariate**.
2. Select these columns:
   - **GDP per capita (constant 2005 US$)**
   - **GNI per capita (constant 2005 US$)**
   - **Adjusted net national income per capita (current US$)**
   - **GNI per capita, PPP (constant 2011 international $)**
   - **Health expenditure per capita (current US$)**
   - **Life expectancy at birth, total (years)**
3. Leave the **Estimation Method** as **Default** and click **OK**.

Figure 11.12 shows the scatterplot matrix for the six columns. In addition to the graphs, the report shows that all of the columns have statistically significant correlations, with the weakest being between adjusted net national income per capita and GNI per capita PPP ($r = 0.1306$. Also note that there appears to be a nonlinear relationship between those two variables. Note further that these variables are measured on quite discrepant scales. For example, the median GDP per capita is \$ 2,316, the median spending on health care is \$ 146, and the median life expectancy is 65.2

years. These order-of-magnitude differences could cause the GDP figures to swamp the others in a subsequent model.

Figure 11.12: Scatterplot Matrix from the Multivariate Report

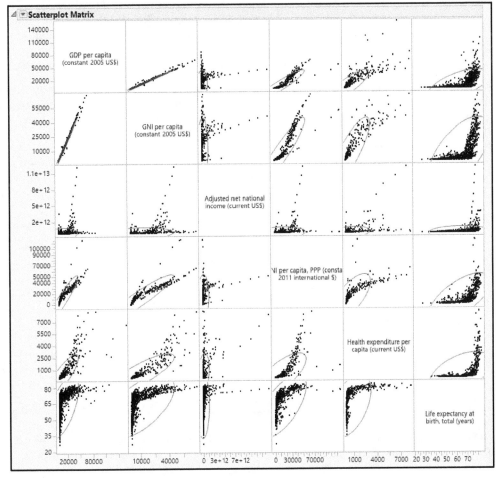

Finally, the report indicates that there are 1,958 missing values in the computation of the correlations. This is quite a large number given that the entire data table has 2,483 rows. By exploring the missing values (not shown here, but left to the reader), we find that we have the most available data for GDP per capita and life expectancy, and that relatively few rows are missing all of the wealth or both of the health measures. As such, PCA should be able to work with the covariance structure of the available data to capture information for father modeling.

1. Now invoke the PCA platform by selecting **Analyze ▶ Multivariate Methods ▶ Principal Components**.

2. Choose the same six columns once more.

3. Leave the **Estimation Method** as **Default** and click **OK**.

The Principal Components platform offers eight different estimation methods, appropriate for data tables with various characteristics. The goal in this illustration is to provide an introduction and a sense of the possibilities that the technique offers. Please consult Chapter 4 in JMP 13 *Multivariate Methods* (SAS 2016) for a full explanation of the options and features of the platform.

As shown at the top of the report in Figure 11.13, the default method in this instance estimates the six principal components using the correlations among variables. Using correlations requires JMP to standardize the data, thereby compensating for differences in the scale of the measurements.

Figure 11.13: Default PCA Report

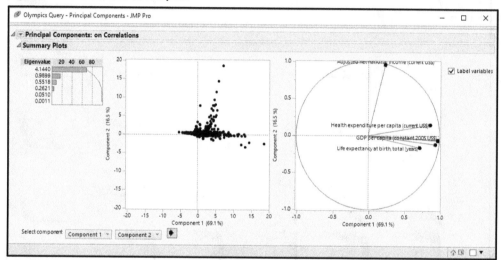

The default report shows three summary plots. The first one to the left lists the eigenvalues and a bar chart with the percent of total variation associated with each principal component. In this case, we see that the first component accounts for slightly more than 69% (69.1%, according to the horizontal axes of the other two plots) of the total variation among the six variables.

Next there is a *score plot*. For each observation, scores are computed for each component based on standardized values. The default score plot has the first two components on the horizontal and vertical axes, respectively. You can select any of the remaining components using the lists at the lower left.

The third graph is a *loadings plot*, which displays the unrotated loading matrix of variables and components. The radius of the red circle is 1.0, and variables that are heavily associated with a component have values closer to 1. Here, for example, life expectancy is the variable most associated with the second component, but also has a weak association with the first. In contrast, the remaining variables all seem to be captured in the first component but to varying degrees.

For more insight into what meaningful content each component has extracted from the original data, we can look at the eigenvalues and eigenvectors that underlie the three plots. Notice that there are dozens of options available. In this chapter, we will demonstrate just a few.

1. Click the red triangle at the upper left of the report.
2. Select **Eigenvalues, Eigenvectors** and **Formatted Loading Matrix**.

Three new elements appear in the report. The eigenvalues report (Figure 11.14) expands on the first summary plot, showing the extent to which each principal component accounts for the total variation, as well as the cumulative percent of variation accounted for. We see that the first two components together account for approximately 86% of all variation, and that the first three account for more than 94%. Going forward, we can reduce the number of columns from six to just two or three for this aspect of the modeling.

Figure 11.14: Eigenvalues Report for the Olympics Data Table

Number	Eigenvalue	Percent	20 40 60 80	Cum Percent
1	4.1440	69.067		69.067
2	0.9899	16.498		85.565
3	0.5518	9.197		94.762
4	0.2621	4.369		99.131
5	0.0510	0.850		99.982
6	0.0011	0.018		100.000

The eigenvectors table (Figure 11.15) reports the coefficients for the linear model that corresponds to each principal component. In other words, the values of each component can be computed through a linear combination of the observed values and the weights listed. The variables are listed in descending order of the coefficients for the first principal component, **Prin1**.

Figure 11.15: Eigenvector Coefficients for the Olympics Data Table

	Prin1	Prin2	Prin3	Prin4	Prin5	Prin6
GDP per capita (constant 2005 US$)	0.48079	-0.06746	-0.19302	-0.16184	-0.43341	0.71624
GNI per capita (constant 2005 US$)	0.48079	-0.06220	-0.19288	-0.15208	-0.46700	-0.69753
Adjusted net national income (current US$)	0.11718	0.96392	0.17219	-0.16292	-0.03059	0.00321
GNI per capita, PPP (constant 2011 international $)	0.46293	-0.11909	-0.02039	-0.52567	0.70309	-0.02078
Health expenditure per capita (current US$)	0.42797	0.14657	-0.32895	0.76934	0.30866	-0.00151
Life expectancy at birth, total (years)	0.35567	-0.16365	0.88728	0.23649	-0.05927	0.00252

Finally, the formatted loading matrix (reproduced in Figure 11.16) is an interactive table of component loadings. These are the values underlying the loadings plot. In this example, we find that all of the variables load substantially into the first component, suggesting that all of these measures capture the same essential construct, which might be thought of as economic well-being. The second component includes the two variables that we originally identified as measures of health, but this report indicates that they provided a small information gain once we've included the wealth measures.

Figure 11.16: Formatted Loading Matrix

⊿ Formatted Loading Matrix	Prin1	Prin2	Prin3	Prin4	Prin5	Prin6
GDP per capita (constant 2005 US$)	0.978735	-0.067117	-0.143384	-0.082860	-0.097904	0.023736
GNI per capita (constant 2005 US$)	0.978732	-0.061882	-0.143278	-0.077862	-0.105492	-0.023116
GNI per capita, PPP (constant 2011 international $)	0.942376	-0.118490	-0.015150	-0.269135	0.158822	-0.000689
Health expenditure per capita (current US$)	0.871214	0.145826	-0.244357	0.393891	0.069723	-0.000050
Life expectancy at birth, total (years)	0.724024	-0.162825	0.659119	0.121079	-0.013389	0.000084
Adjusted net national income (current US$)	0.238534	0.959028	0.127910	-0.083412	-0.006911	0.000106

Suppress Absolute Loading Value Less Than   [0.3] ⊏━━◇━━━━━━▻

Dim Text   [0.4] ⊏━━◇━━━━━━▻

The opacity of the numbers displayed varies with their absolute values. Below the table are sliders that give the user control over which values are dark.

Finally, we can construct and save as many of the components as we deem useful for further molding purposes. We can also save components with imputation for missing values, potentially using all available data to construct components for each row in the original table.

1. Click the red triangle once more, and choose **Save Principal Components**.

2. In the dialog that opens, enter the number of components you want to save. We'll reason from the loadings that three components sensibly represent the original six variables adequately, so will change the suggested number to 3. This adds three new columns to our data table, which are grouped together as Principal Components (3/0). Before discussing them further, let's also save the components with imputation of missing values.

3. Again, click the red triangle; choose **Save Principal Components with Imputation**; and again specify three components. Three new columns are added to the data table, several rows of which appear in Figure 11.17.

Figure 11.17: Comparing Principal Components with and without Imputation

			opulation, total	Prin1	Prin2	Prin3	Prin1 2	Prin2 2	Prin3 2
▾ Olympics Query									
SQL SELECT t4."Country Name", t4."Country Code", t4.Year, t1.C									
▷ Source		13	26528741	•	•		•	-0.94972616	0.13326034...	0.01071295...
▷ Load pictures for "National Olympic Committee"		14	1608800	•	•		•	-0.0227499...	0.00250051...	-0.0075575...
▷ Modify Query		15	1814135	•	•		•	0.370109123	-0.0406797...	0.12295052...
▷ Update From Database		16	2022272	•	•		•	0.49803654...	-0.0547406...	0.16544810...
		17	2243126	•	•		•	0.63512506...	-0.0698084...	0.21098901...
		18	2458526	•	•		•	0.82386067...	-0.0905529...	0.27368712...
▾ Columns (32/0)		19	2671997	•	•		•	-0.39368511	-0.1471088...	0.83027268...
⊿ GNI per capita, PPP (constant 2011 international $)		20	2904429	•	•		•	-0.3898391...	-0.0710809...	0.902762822
⊿ GNI, Atlas method (current US$)		21	3142336	•	•		•	-0.3732486...	-0.0803600...	0.96051390...
⊿ GNI, PPP (constant 2011 international $)		22	3247089	•	•		•	-0.4493211...	-0.0771690...	0.98480727...
⊿ Gross capital formation (current US$)		23	3168033	-0.3958271...	-0.0894982...	1.02337000...	-0.3958271...	-0.0894982...	1.02337000...
⊿ Gross enrolment ratio, primary, gender parity index (GPI)		24	3089027	-0.2964484...	-0.1218714...	1.14252598...	-0.2964484...	-0.1218714...	1.14252598...
⊿ Gross enrolment ratio, secondary, gender parity index (GPI)		25	3026939	-0.1443504...	-0.1402627...	1.21975902...	-0.1443504...	-0.1402627...	1.21975902...
⊿ Health expenditure per capita (current US$)		26	2947314	0.01757921...	-0.1471892...	1.23068985...	0.01757921...	-0.1471892...	1.23068985...
⊿ Life expectancy at birth, female (years)		27	11124892	•	•		•	-1.1827142...	0.27544603...	-0.9796413...
⊿ Life expectancy at birth, male (years)		28	12295973	•	•		•	-1.1528958...	0.22434514...	-0.8334865...
⊿ Life expectancy at birth, total (years)		29	13744383	•	•		•	-1.08199372	0.19585995...	-0.7176262...
⊿ Percentage of students in secondary education who are fer		30	15377095	•	•		•	-0.9496785...	0.210479904	-0.5986696...
⊿ Percentage of students in secondary general education wh		31	17190236	•	•		•	-0.8421962...	0.18916514...	-0.4257748...
⊿ Population, total									
▷ Principal Components (3/0)									
▷ Principal Components 2 (3/0)									

Rows 15 through 28 correspond to Albania, which had missing values in varying years for the wealth variables, only five years of health expenditure data, but life expectancies for all observed years. There was a full complement of six variables observed for the final five years of the time period. Hence, components could be computed for those five years. For the prior years, the algorithm uses those features that were available in combination with the component coefficients to estimate component values for the remaining years.

It is important to remember that PCA is a linear method, so it does not capture the nonlinearity that we noticed earlier with respect to life expectancy and health expenditures. We might do better with a regression modeling approach to combine those two variables and impute the missing observations.

For a more comprehensive explanation of PCA, readers might refer to Chapter 4 of Shmueli *et al.* (2016) and Chapter 4 in JMP *Multivariate Methods* (2016).

Abundance of Rows

It might seem counter-intuitive to suggest that there are times when we might want to reduce our sample size. Traditionally, statisticians seek larger rather than smaller samples. Particularly with data mining or machine learning projects or with studies that use data sets with large numbers of observations, there are at least three scenarios in which it might be advantageous or desirable to build models with fewer rather than more rows:

- In general, we don't want to develop hypotheses and then test them with the same data. When we have the luxury of an enormous number of observations, we can productively use part of the data for model-building, and then validate the models with a holdout sample. A standard practice in data mining is to divide a body of data into training and test subsamples. In JMP, we can easily partition a large sample for exploration, modeling, and validation by creating a two- or three-level dummy validation variable. Many analysis platforms support the use of a validation column.
- If the observational unit is at a finer level than the modeling task demands, or if we have data from multiple sources that use different observational units, we might want to aggregate rows in some fashion. For example, we might find some city-level data when we want county- or state-level data, or we have repeated measures that we want to combine for each individual.
- When building models to predict rare events, we might want to over-sample the rare event and under-sample others. Hence, we might randomly suppress rows corresponding to the unusual target values.

In this section, we briefly explain and demonstrate how these row-reduction tasks can be accomplished. As with most of the examples in the chapter, these examples are intended to point the reader in a direction rather than serve as a comprehensive and complete treatment of the subject.

Partitioning into Training, Validation, and Test Sets

It is *de rigueur* in data mining projects to randomly split a large sample into at least two subsets: a *training set* to select features (variables) and choose among available modeling methods, and a *test set* to confirm model suitability and avoid over-fitting. Some authorities (for example,

Truxillo 2012) recommend splitting into three partitions to develop and validate the model, and finally testing its performance on a third clean set of data for honest assessment.

It is common to find reference to training and test data sets within the literature on data mining. In the SAS world, the first—and typically larger—split is also called the training set. The second set is referred to as the validation set, and the third partition would be the test set. As with so much terminology in this area of analytics, the language is evolving and hence can be confusing.

With time series data, it is better to split the data nonrandomly by building the model with earlier rows, and then testing with a holdout sample of most recent observations. This issue is briefly described in a later section of the chapter.

We demonstrate partitioning with the **Diamonds** data in the **JMP Sample Data** directory and using the **Fit Model** platform to illustrate the introduction of a validation variable. This data set contains attributes and prices for a sample of 2,690 diamonds. The sample data table includes several saved scripts that will facilitate our demonstration. Open the data table now.

Rather than fit a model using the entire set of observations, let's first define training and test sets by creating a new dummy variable.

- Select **Cols ▶ New Columns**. Complete the dialog as shown in Figure 11.18, editing the areas highlighted in the figure.

This will create a new column called **Training**, consisting of 0s and 1s as the first column in the data table. The training set (the 0s) consists of 80% of the rows, and the remaining 20% are the test set.

Figure 11.18: Defining a Column to Split a Data Table into Training and Test Sets

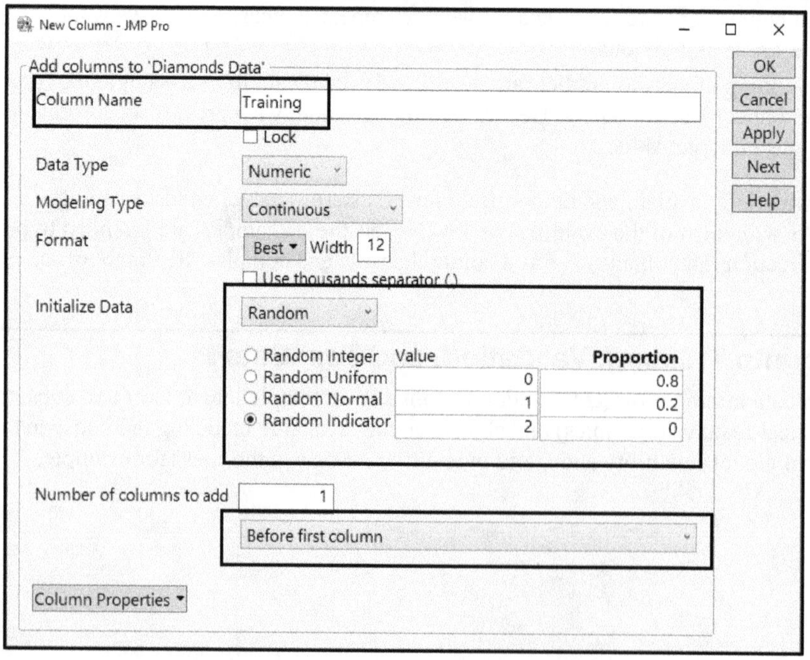

For some operations, such as exploratory graphing or descriptive statistics, you can simply hide and exclude those rows where **Training = 1**. For other more involved analyses, we can use the new column to train a model and compare its performance to the same model with the test subsample.

The **Diamonds** sample data has several scripts saved within it. Let's launch the **Model** script and amend it by including the validation column in the mix. (See Figure 11.19.) This script specifies a multiple regression model for the **Price** variable in the table.

1. Click the green arrowhead next to Model in the upper left of the data window.
2. In the **Fit Model** launch window that opens, take a moment to review the **Model Specification**. The target (**Y**) variable is **Price**, and the model includes seven potential features.
3. Before running the model, highlight **Training** under **Select Columns**, and click **Validation**. This tells JMP that the column called Training should be used to split the original sample, and then to report the model's goodness of fit statistics for the training and validation data. Then **Run** the model.

Figure 11.19: Fit Model Platform including a Validation Column

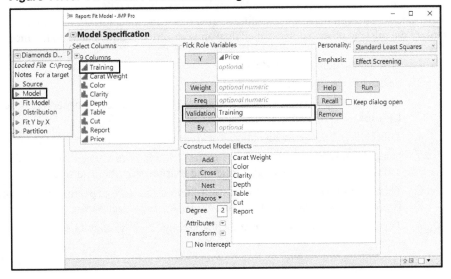

4. Within the report window, scroll down until you find the **Crossvalidation** report (Figure 11.20).

Figure 11.20: The Crossvalidation Report

⊿ Crossvalidation			
Source	**RSquare**	**RASE**	**Freq**
Training Set	0.9175	694.98	2152
Validation Set	0.9138	711.18	538

JMP estimates the multiple coefficients just using the 2,152 training observations. It then computes predicted values and residuals for the entire original sample based on these coefficients, and reports the unadjusted **RSquare** and the Root Average Squared Prediction Errors (**RASE**) for both subsamples. Had we created the **Training** column with three values rather than two, this report would list Training Set, Validation Set, and Test Set.

These summary statistics can indicate how well a particular model specification performs on data outside the training set. In practice, we might fit several models and ultimately select one that performs well both on the training set and the validation set. This can avoid the problem of overfitting and produce a model so attuned to the training data that it does not generalize outside of the original sample.

Aggregating Rows with Summary Tables

In some studies, we might want to work with summary statistics of individual rows rather than the original rowwise observations to meet the goals of a study. For example, in the Olympic medals data, it might be sensible to look at the overall total number of medals won by countries rather than the number of medals collected in a given year. In the diamonds data, perhaps we want to look at the mean prices of diamonds aggregated by color. Doing so can substantially reduce the number of rows.

Let's return to the Olympics data table to create a reduced data table with the name of the country, total medals won over all 14 observed Olympic games, mean GDP per capita, mean life expectancy, and mean population. Before summarizing the medal winnings by country, we'll replace the missing observations of medals won with zeros, and then sum up the medals.

1. Select the column **TotMedals**.
2. Select **Cols ▶ Recode**.
3. There are 2002 missing values (.). Recode them to **0**, click **Done**, and choose **New Column**.
4. This creates a new column called **TotMedals2** as the last column in the table.
5. Select **Tables ▶ Summary**. (See Figure 11.21; in prior versions of JMP, this was called **Tabulate**). Select **TotMedals2**, click the **Statistics** button, and select **Sum**.
6. Then select **GDP per capita (constant 2005 US$), Life expectancy at birth, total (years)**, and **Population** total. Click **Statistics** and select **Mean**.
7. Then choose **Country Name** and click **Group**. Finally, click **OK**.

Figure 11.21: Summary Launch Window

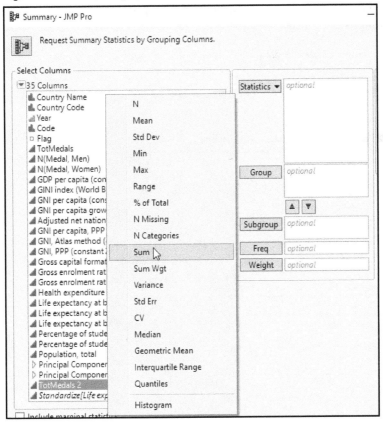

This creates a new data table with the five columns that we selected, plus a sixth column showing the number of rows summarized from the original table. This new table has merely 191 rows, rather than the original 2,483 rows. Each row in the new table represents a single country with values summed over all Olympic years, rather than a single country in one year, as in the main table.

Oversampling Rare Events

When the dependent variable occurs rarely within a data set with a very large number of rows, it is conceivable that a randomly selected training sample will have too few observations of the event of interest to provide much power. In such cases, it can be useful to construct a stratified random sample where the strata are defined by the presence or absence of the rare event.

To illustrate the approach, let's return again to the **Health Risk Survey** data in the JMP Sample Data directory. Recall that the data table contains health and lifestyle data gathered from almost 14,000 teens. One of the **Dichotomous Response Questions** asked whether a respondent had injected IV drugs at least once in their lives. Only 282, or 2% of those responding answered affirmatively. There were 120 missing observations.

If we wanted to build a classification model to understand factors associated with this particular behavior, it would be important that the training sample have a moderate number of Yes responses. The following steps show one way to do just that.

1. Open the **Health Risk Survey** data from the sample data directory.
2. Select **Tables ▶ Subset**. (See Figure 11.22, below.)
3. Click the button next to **Random – sample size:** and leave the default value of **1000** in place.

Figure 11.22: Subset Tables Launch Window

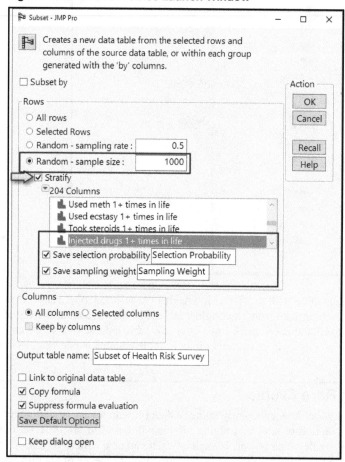

4. Check the box next to **Stratify**.
   a. Next, under 204 Columns, scroll down to the group called Dichotomous Response Questions. Expand the group list by clicking the gray arrow.
   b. Scroll until you find the column **Injected drugs 1+ times in life** and select it.

   In a stratified subset defined by sample size, JMP will attempt to select the specified number of observations from each stratum. If any stratum has fewer observations, it will select all of the rows in that stratum.

5. Check the boxes marked **Save selection probability** and **Save sampling weights**, and click **OK**. This will enable you to apply analysis weights at a later time if that suits the purposes of the investigation.

A new data table will open containing all of the columns and scripts from the original table, but with 1,402 rows. The rows include all 282 teens who responded Yes to the IV drug question, plus the 120 teens who did not respond, and an additional randomly chosen 1,000 teens who responded No. We could now proceed to create a validation column as shown earlier and do our modeling and other analysis.

Date and Time-Related Issues

In JMP, as in other software, times and dates are stored internally as numbers, but might be displayed in a variety of more recognizable forms. JMP stores date-based data internally as the number of seconds elapsed since January 1, 1904, and provides numerous formatting options as well as date and time functions to carry out the special operations demanded by date data.

Working with data and time data can present its own special issues, such as these:

- Although time is continuous, ordinary arithmetic operations don't apply.
- JMP uses the 1904 bases, but not all programs do the same. For example, some versions of Excel use 1904, but others use 1900 as the starting point. IBM SPSS uses October 14, 1582, and UNIX systems use January 1, 1970. For projects drawing data from various originating sources, it could be necessary to take these differences into account.
- Sometimes, we might want to extract a part of a timestamp—for example the day of the week from a date, or just the hour from a time.
- Similarly, we might want to aggregate rows across time, which could require extraction of (for example) the year from dates expressed as mm/dd/yyyy.
- For observations that have a time associated with them, sometimes we might be less interested in a specific observation than in its change from a prior value or in the prior value itself.

This section surveys some of these issues and illustrates starting points for further exploration of JMP date-handling capabilities. Because every project will have its own unique demands, it is unrealistic to provide a comprehensive treatment of the subject in a few pages. Readers are referred to the JMP documentation for further details about the techniques shown here.

- Open the JMP sample data table called **Stock Prices**, which contains daily price information for a particular stock for the period from October 7 through December 30, 2011.

Formatting Dates and Times

The first column in this data table contains the date, formatted as d/m/y. We can customize that format using the **Column Info** capability.

1. In the **Columns** panel, position the pointer over the word **Date** and right-click. Choose **Column Info**. Note that the variable is **Numeric** and **Continuous**.

2. As shown in Figure 11.23, click the **Format** marked **d/m/y** and choose **Time**. This opens another menu of twenty date and time formats. Select the one that suits the purpose of the analysis.

Figure 11.23: Data and Time Format Options

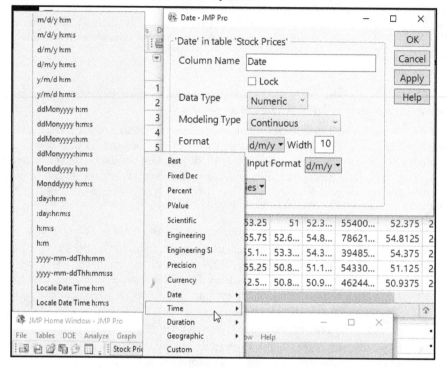

Changing the format does not change the internally stored value of the date but can facilitate interpretation of graphs and reports.

Some Date Functions: Extracting Parts

JMP has numerous functions for working with time-related columns, particularly with respect to isolating portions of a time stamp. In the current example, you might want to focus on the month or quarter of a date. In another setting, you might be interested in the hour of the day, or day of the week.

In the stock price data, move into the **Columns** panel, and scroll to the final column (**YearWeek**). Then click the large plus sign next to it to open the **Formula Editor**. This column is computed from the **Date** column, and nicely illustrates one of the many ways that date functions can be used. Figure 11.24 shows the formula, as well as many of the date and time functions available.

Figure 11.24: Creating a New Variable from a Date

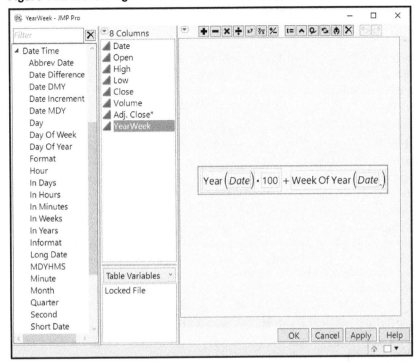

Another interesting data table demonstrating several time-related functions is found on the JMP User Community website. It contains readings from two solar arrays on the SAS Campus in Cary NC (Gregg 2014). Sensors recorded data every 15 minutes for eight months in 2009, and JMP developer Xan Gregg shared this data table that includes five computed columns representing observation times in several numeric and categorical ways.

Aggregation

In an earlier section of this chapter, we learned how to aggregate rows by country in the Olympic medal data. We can also aggregate across time or date. For example, we might want to summarize the stock price data by the week, perhaps computing the mean closing price or total (summerized) volume for each week. The operation is the same as that shown earlier (Figure 11.21), but using the relevant date column for grouping.

Row Functions Especially Useful in Time-Ordered Data

Earlier in the chapter, we saw methods for taming skewed distributions, including computing the log of a variable. With data gathered over time, we can sometimes achieve the same goal by computing the *difference* between the value in row k and the prior value. Alternatively, for some modeling purposes the *lagged*, or prior, value might be more informative than the current value.

We'll demonstrate both using **Graph Builder**. You could also create new data table columns using the **Formula Editor**, but this approach more readily illustrates what these transformations offer.

1. With the **Stock Prices** data table open, launch **Graph Builder**.

2. Drag **Volume** to the **X** drop zone and click **Histogram**. Daily trading volumes are strongly right skewed.

3. Now, position the pointer over **Volume** in the list of columns in the Variables panel of the **Graph Builder** launch window and right-click.

4. Select **Row ▶ Difference**. This adds a new item, *Difference[Volume]*, to the list of columns.

5. Drop *Difference[Volume]* in the **X** drop zone to replace **Volume**. Note that the new histogram is quite symmetric.

   Differences use the lagged, or prior row, value to calculate a daily change in stock prices. In some modeling, we might want to refer to the lagged value itself. For example, the so-called *Random Walk* model of stock prices asserts that we can use yesterday's price to forecast today's price, or that today's price will be a small random departure from yesterday's price. In other words, the model hypothesizes that the closing price on day *t* is a linear combination of the prior day's price plus a random error term. Let's see the random walk at work.

6. In **Graph Builder**, click **Start Over**.

7. Drag **Adj. Close*** to the **Y** drop zone, and **Date** to the **X** zone. There is a downward trend, with day-to-day ups and downs.

8. Position the pointer over **Adj. Close*** in the list of columns and right-click.

9. Select **Row ▶ Lag** to adds *Lag[Adj. Close*]* to the list of columns.

10. In the **X** drop zone, replace **Date** with *Lag[Adj. Close*]*.

We now have a reasonably linear scatterplot, suitable for a simple linear fit.

Elapsed Time and Date Arithmetic

Finally, it is sometimes useful to compute differences between time stamps or to add or subtract a constant to a date or time variable. For example, if observations include the start and ending times of an event, we might want to find the elapsed time. If we have birthdate in a column, we might want to compute age. In another context, we might want to calculate a future date—for example, 30 days following a transaction.

In addition to the JMP functions for extracting or re-expressing dates, there are two specialized functions called Date Difference and Date Increment. Date Difference computes the difference between two date-time variables. Date Increment adds a fixed time interval to a date-time variable. Interested readers should consult the Formula Functions Reference, Appendix A, in *Using JMP 13* (SAS 2016).

Conclusion

Before performing analysis or training models, there are a variety of penultimate preprocessing steps that might be needed. This chapter has presented some common scenarios and illustrated a selection of approaches that can meet those needs. In addition, it has directed readers to resources in the JMP documentation for further particulars.

At this point, our discussion of data management and the dirty work of analytics is nearly complete. All that remains is a very brief discussion of ways to export a clean data table for users who do not use JMP. That is the topic of Chapter 12.

References

Gregg, Xan. 2014. "SAS Solar Array Data Jan – Aug 2009." *JMP Sample Data.* Downloaded from https://community.jmp.com/t5/JMP-Sample-Data/SAS-Solar-Array-Data-Jan-Aug-2009/ta-p/21427, December 28, 2016.

Han, Jiawei, Micheline Kamber, and Jian Pei. 2011. *Data Mining: Concepts and Techniques, Third Edition.* Waltham MA: Morgan Kaufmann, Chapter 3.

IBM Knowledge Center. 2016. "Dates and Times in IBM SPSS Statistics." Downloaded from http://www.ibm.com/support/knowledgecenter/SSLVMB_21.0.0/com.ibm.spss.statistics.help/idh_idd_dtwz_learn.htm, 28 December, 2016.

Kandel, Sean, and Jeffrey Heer, Catherine Plaisant, Jessie Kennedy, Frank van Ham, Nathalie Henry Riche, Chris Weaver, Bongshin Lee, Dominique Brodbeck, and Paolo Buono. 2011. "Research directions in data wrangling: Visualizations and transformations for usable and credible data". *Information Visualization.* Vol 10(4), pp. 271-288.

Kuhn, Max. and Kjell Johnson. 2013. *Applied Predictive Modeling*, Chapter 3. New York: Springer.

McCormack, Don. 2015. "It's a dirty job, but someone has to do it." Presentation at *JMP Discovery Summit 2015*. Downloaded from https://community.jmp.com/t5/Discovery-Summit-2015/It-s-a-Dirty-Job-but-Someone-Has-to-Do-It/ta-p/23806.

SAS Institute Inc. 2016. JMP 13 *Basic Analysis*, Chapter 3. Cary NC: SAS Institute Inc.

SAS Institute Inc. 2016. JMP 13 *Multivariate Methods*, Chapter 4. Cary NC: SAS Institute Inc.

SAS Institute Inc. 2016. JMP 13 *Predictive and Specialized Modeling*, Chapter 15. Cary NC: SAS Institute Inc.

SAS Institute Inc. 2016. *Using JMP 13*. Cary NC: SAS Institute Inc.

Shmueli, Galit, and Peter C. Bruce, Mia L. Stephens, and Nitin R. Patel. 2016. *Data Mining for Business Analytics: Concepts, Techniques, and Applications with JMP Pro.* Hoboken, NJ: John Wiley, Chapter 4.

Son, Nguyen Hung. No date. "Data cleaning and data preprocessing." Downloaded from http://duch.mimuw.edu.pl/~son/datamining/DM/4-preprocess.pdf, December 6, 2016.

Svolba, Gerhard. 2006. *Data Preparation for Analytics Using SAS.* Cary NC: SAS Institute Inc.

Truxillo, Catherine. 2012. *Advanced Business Analytics Course Notes.* Chapter 2. Cary NC: SAS Institute Inc.

Chapter 12: Exporting Work to Other Platforms

Introduction

In the context of the entire ongoing process of analytic inquiry, this book has focused extensively on the steps that precede modeling and analysis. The operating assumption is that the analyst intends to use JMP exclusively for the analysis phase. This book is decidedly not about actually fitting models, testing hypotheses, or making predictions. Those topics are amply treated in numerous other sources.

There are, however, phases of a project following analysis, and within most organizations there are other tools for modeling and reporting. In this chapter, we sketch some methods for exporting a clean JMP data table to another software environment as well as some ways to integrate JMP reporting with other popular platforms.

Why Export or Exchange Data?

Just as JMP users frequently need to import raw data from other software environments, we might also want to transfer JMP data that are formatted for use by other packages. Owing to organizational standards, client needs, regulatory requirements, or specialized modeling purposes, you might not complete an entire project exclusively using JMP.

It was clear in earlier chapters that JMP anticipates the need to read data in a wide variety of native formats. Unfortunately, other analytic software might not reciprocate with an ability to read

*.jmp files. Depending on the destination software, though, JMP can prepare a data table to be shared in other environments.

Toward the end of the chapter, we'll also review some of the ways to share the results of an investigation with JMP users and non-users.

Fit the Method to the Purpose

If your goal is to read a JMP data table using SAS, no additional steps are necessary. Using PROC IMPORT and the dbms= JMP replace option, SAS has the ability to read data that were written in JMP. Alternatively, JMP can save a file in a format that is compatible with SAS. For other environments, there are two basic ways to proceed. You can either *save* the data in a format that the other program will accommodate (text, CSV, and so on) or export the data directly to a SAS or database server, assuming that you have appropriate permissions. Most of these avenues are accessible from the JMP **File** menu.

Save As

This essentially is the complement of the techniques used to read a single flat file at a time, and requires minimal explanation. We'll illustrate using the **Seasonal Flu** data found in the **Sample Data** directory. Open that file.

1. Select **File ▶ Save As**.
2. Navigate to the directory in which you want to save the exported data.
3. At the bottom of the dialog, click the drop-down list next to **Save as type**. Figure 12.1 shows the available options. Choose one that the destination software can read.

Figure 12.1: File Format Options in Save As

File name:	Seasonal Flu
Save as type:	JMP Data Table (*.jmp)

JMP Data Table (*.jmp)
Excel Workbook (*.xlsx;*.xls)
Text Export File (*.txt)
JSON Data File (*.json)
CSV (Comma delimited) (*.csv)
TSV (Tab delimited) (*.tsv)
SAS Data Set (*.sas7bdat)
SAS Transport File (*.xpt)
dBASE Files (*.dbf;*.ndx;*.mdx)

ide Folders

A JMP data table contains both data and metadata. The alternate formats will all save the data, but not necessarily formats, data types, variable names, data labels, or formulas. For example, SAS will not honor the data type if you specify variables as nominal, ordinal, or continuous. After you select the desired format and proceed to save the file, JMP presents an alert (Figure 12.2).

Figure 12.2: File Save As Alert Message

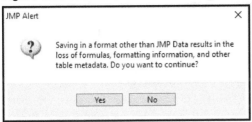

It is incumbent on the analyst to investigate the capabilities of the destination software and select the option that preserves the most information from the original JMP table. It is worth noting that the SAS Data Set (*.sas7bdat) option has the greatest capability and flexibility in preserving metadata. If the destination software will read a SAS file, that is your best choice.

If you are exporting to Excel, be aware that you can save multiple data tables as worksheets within a single Excel workbook. Open the related data tables, and select **View ▶ Create Excel Workbook**. Then follow the steps in the dialog. For explicit instructions, see Chapter 4 of *Using JMP 13* (SAS 2016).

Whichever option you select, you will be saving the data file in a local or network directory accessible from your desktop. You can also save data to a database with a corresponding ODBC driver or to a SAS Library, as outlined in the next section. In this way, you actually push the local JMP file to the receiving system.

Export to a Database

Chapter 5 explained how to query a database and build a JMP data table in the process. We can also reverse the process in a limited way. As before, it is necessary to have a database connection with appropriate authorization and passwords. If you have no current database connections, skip to the next section.

1. Open the table or tables that you want to export.
2. Select **File ▶ Database ▶ Save Table**.
3. This opens a launch window asking you to select a connection and the schema into which to save the new table. See Figure 12.3.

Figure 12.3: Saving JMP Data Tables to a Database

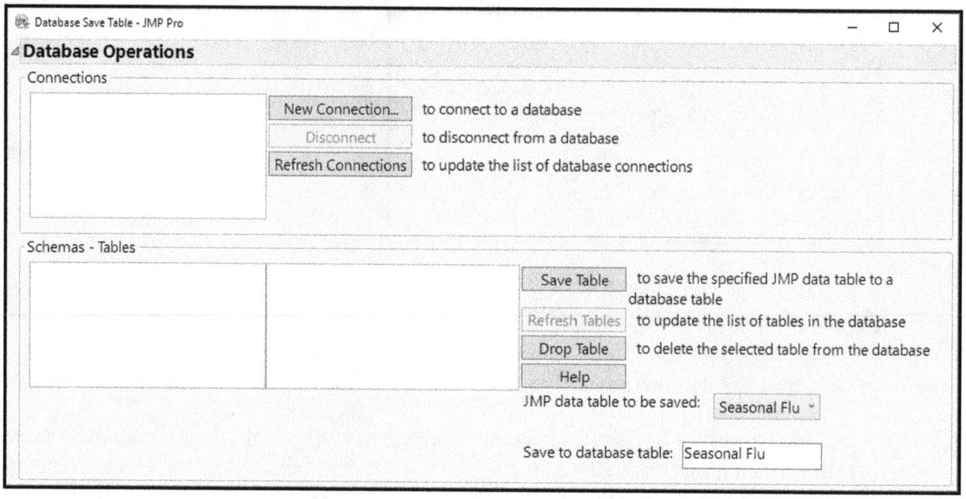

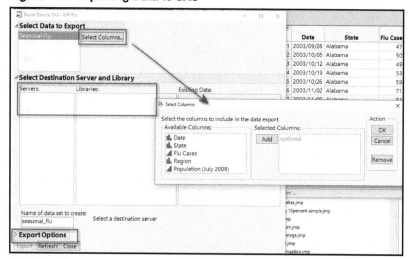

4. Choose from among your available connections, select the appropriate schema, and specify the one or more tables that you want to save.

For further examples and illustration, see "Save Data Tables to a Database" in *Using JMP 13*.

Export to a SAS Library

Here again, if you have no access to SAS Libraries, move past this section. For users wanting to pass a clean JMP table to SAS for further analysis, we've seen above how you can save data in SAS format, and then switch to SAS to import it. Alternatively, workflow can be expedited to directly save the JMP file to the relevant SAS library. As an added advantage, the method permits the user to select columns to export, so that you do not need to export an entire data table.

1. Select **File ▶ SAS ▶ Export Data to SAS**. This opens the launch window in Figure 12.4.

Figure 12.4: Exporting Data to SAS

2. In the upper left, choose the data table to export.
3. If you want to export only some columns, click **Select Columns** and complete the dialog as desired.
4. In the central portion of the launch window, choose the destination server and library.
5. At the lower right, note that you can also change the name of the data file as well as make optional choices related to excluded rows, SAS variable names, and SAS formats.
6. When you are finished, click **Export**.

You should then see a JMP alert telling you that the export was successful, and you should find that there is now a new *.sas7bdat file in the specified SAS library. You can then access the file from SAS.

Complete documentation and another example is available in Chapter 3 of *Using JMP 13*, under "Export JMP Data Tables to SAS."

Exporting Reports

Experienced JMP users might be familiar with JMP journals, projects, and reports. These are valuable for documenting work on a project and are easily shared with other JMP users. As we have seen, we can preserve a record of all ETL operations so that work can be reproduced at a future time.

Although the focus in this book has been on the pre-analysis stages of a project, it is worth noting that JMP has continued to enhance its functionality for sharing reports and graphics, including interactive graphics, *outside* of the JMP environment. This means that an analyst can prepare a written report, a PowerPoint presentation, or a website directly from a JMP session. This section surveys a few of the ways that JMP products can be passed to the outside world. Some options preserve the data, animation and full interactivity. But others essentially take a snapshot of a current status of the visualization. Similarly, for reports containing tables or formatted text, the options differ in terms of the extent to which formats are preserved.

Since the **Seasonal Flu** data table includes two scripts that create interactive bubble plots, let's begin with the first of them and describe major options for export. Click the first green arrow to run the script creating a bubble plot of flu cases by date, sized by population (Figure 12.5). The methods shown in the following discussion also apply to other interactive reports in JMP, such as profilers.

Figure 12.5: First Seasonal Flu Bubble Plot

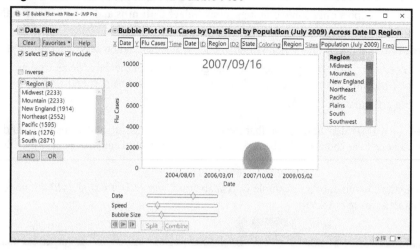

This graph is interactive and features animation. Within JMP, a viewer can swap out variable roles, adjust the speed and size of the bubbles, and otherwise investigate the data. The user can select or de-select multiple regions with the local Data Filter. If we want a user without JMP to have access to this work product, we have several alternatives. These options differ slightly between Windows and Mac versions of JMP. See "Save and Share Reports" in Chapter 10 of *Using JMP 13* for full details.

Interactive Graphics

For maximally retained interactivity, you can save a bubble plot or some other interactive report either as Interactive HTML or as an Adobe Shockwave Flash file. In both options, the graphic as well as the underlying data are embedded in the file, and some (if not all) interactive elements are functional.

The availability of HTML5 support and the ability to create Shockwave files has increased during recent releases of JMP. Readers should consult the documentation for their own particular versions, but this is a general guide:

HTML5 Support:

- JMP 11: Tables and many common Graphs
- JMP 12: Support added for Bubble Plots, Profilers, and tablets
- JMP 13: Support added for most features of Graph Builder

Adobe Shockwave Flash file creation for Bubble Plots, profilers created from the Graph menu, and Histograms in the Distribution platform.

To save as an HTML file, select **File ▶ Save As**, and choose **Save as Type: Interactive HTML (.htm, .html)**. Be aware that there is a related choice (**Hypertext Markup Language (.htm, .html)**) that will save the text and references to image files, which are saved separately. There is no interactivity with this option. The availability of these options varies across JMP platforms, so users should consult JMP documentation for the version of JMP in use.

For Bubble Charts, interactivity is only partially implemented, So this option creates an HTM file so that users can explore data, brush regions of a graph, and so on. For a clearer sense of the capabilities, try this:

1. Select **Analyze ▶ Distribution**. Select **Flu Cases**, **Region**, and **Population (July 2009)**. Click **OK**.
2. Select **File ▶ Save As**. Name the output file as you wish, and be sure to choose **Save as Type: Interactive HTML (.htm, .html)**.

Your default browser will now open with the **Distributions** report. You can brush any area of one of the graphs, and the corresponding rows in the other graphs will be selected. You can close or reveal individual parts of the report as well. The web page includes some controls provided by JMP as depicted in Figure 12.6.

Figure 12.6: Controls in the Interactive HTML File Display

A third interactive option—at least for bubble plots—is to create an Adobe Shockwave Flash file. This file that can be displayed in a web browser and can also be embedded in other files such as PowerPoint presentations. This option retains nearly all of the interactivity and animation, so we'll discuss it in the context of the bubble plot.

This is not accessed from the **File** menu, but rather from the red triangle at the top of the report. In this case, it is the hotspot next to **Bubble Plot of Flu Cases**. Among the options listed on the pop-up menu is **Save for Adobe Flash platform (.SWF)**. You will be prompted for a directory and filename once again. After you save, your default browser will open, this time displaying a fully interactive object reproduced in Figure 12.7.

Figure 12.7: Interactive Bubble Plot as Flash File

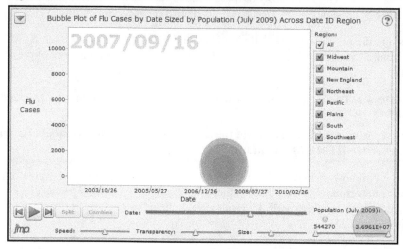

Finally, a fourth option is a *web report*. As described in *Using JMP 13*, this option "creates a web page in which reports, descriptive text, and graphics are displayed. The web page, graphics, and support files are saved in the directory that you specify so that you can zip the files and send them to another user. This feature is particularly helpful for non JMP users" (SAS 2016, p. 429).

Working from the bubble plot once again, select **View ▶ Create Web Report**. In the launch window, select the reports to include, a name for the folder that will hold all of the components, as well as a destination directory. Further customization of titles and descriptive text is allowed, and in this way you can package multiple HTML interactive files along with the supporting data.

Static Images: Graphics Formats, PowerPoint, and Word

Interactivity is an extremely valuable feature of many JMP reports, but sometimes your reporting goals are fully served by static images and tables. We won't illustrate all of the available options here, but we want to call attention to some. They are all to be found on the **File** menu. The full list of file types appears in Figure 12.8. Be aware that these options are available to Windows users; Macintosh users have more limited options under **File ▶ Export**.

Figure 12.8: Save As Options for JMP Output

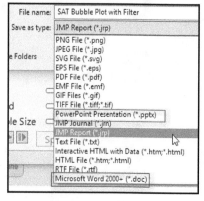

The first group of eight options are mostly graphical formats, with the exception of the Portable Data Format (PDF), and they all produce essentially static pictures of a particular JMP report. Graphics can be saved at various resolutions, which can be important for in scientific publications that might require high resolutions (for example, 300 dpi).

We call special attention to the two highlighted options in Figure 12.8. Any JMP report can be exported as either a PowerPoint presentation or a Microsoft Word document. JMP provides a standard template for both, and in both cases pictures, text, and tables are preserved. Once the files are created, they can be edited and modified within PowerPoint and Word as needed.

Conclusion

Even ardent users of JMP will need to share data and reports with users who don't use SAS inside and outside of their organizations. This chapter has illustrated some of the ways to do so, and should provide a foundation for analysts to share portions of their projects efficiently across platforms.

References

SAS Institute Inc. 2016. *Using JMP 13,* Cary NC: SAS Institute Inc.

Index

Ready to take your SAS® and JMP® skills up a notch?

Be among the first to know about new books, special events, and exclusive discounts.
support.sas.com/newbooks

Share your expertise. Write a book with SAS.
support.sas.com/publish

sas.com/books
for additional books and resources.

THE POWER TO KNOW.

Printed in the USA
CPSIA information can be obtained
at www.ICGtesting.com
LVHW080722141123
763836LV00015BA/85